This book is dedicated to all scientists and researchers, who are working in the search for truth and a better understanding of the world.

Heinz D. Knoell *received his B.A., B.Sc., M.Sc., and Ph.D. at the University of Muenster, Germany. From 1980 on he worked as an IT-Consultant and from 1986 on was for nearly 30 years professor of Computer Information Systems at the University of Applied Sciences in Lueneburg, Germany (now merged with Leuphana University).*
From 1978 on, he was on many business trips to several Universities in the USSR and Russia.
Since his retirement he studies Psychology and Soviet History at the University of Hamburg, Germany.

Jerwen Jou *received his B.A. at the National Chengchi University, Taiwan, his M.A. at Fujen University, Taiwan, M.S. in Psychology at East Texas State University, and his Ph.D. in Psychology at Kansas State University. Since 1990 he is with the University of Texas Rio Grande Valley in Edinburg, TX, since 2000 as professor of Psychology.*

Heinz D. Knoell
Jerwen Jou

Russian and Soviet Psychology in the Changing Political Environment

A Time Series Analysis Approach

tredition®
Hamburg, Germany

Translation of Russian and German references and excerpts in books and journals into English by Heinz D. Knoell

Publisher & Print: tredition GmbH,
Halenreie 40-44,
22359 Hamburg
Germany

ISBN
978-3-347-30834-3 (Paperback)
978-3-347-30835-0 (Hardcover)
978-3-347-30836-7 (e-Book)

Contents

Preface

This monograph describes psychology in changing political environments in Tsarist Russia in the mid of 19th century, the Soviet Union, and the Russian Federation until the year 2000. Russia and the Soviet Union are of special interest, because of the multitude of political changes. There were not only the so-called October-Revolution in 1917, which initiated the Soviet rule, and the breakdown of the Soviet Union in 1990. But there were also a multitude of changes in Soviet times. In the beginning of the Soviet rule there were many new developments in sciences, which also affected psychology. Since Stalin assumed power, there were many developments and ideological restrictions as well. These are well documented in books and journal articles (e.g. Artemyeva, 2015; Bauer, 1959; Ehrsam, 1985; Graham, 1972; Kozulin, 1984; Krementsov, 1997; Kussmann, 1974; McLeish, 1975; Petrovsky, 2000; Thielen, 1984).

At first glance it may seem that after Stalin's death there was a stable ideology, which guided the sciences and research. This was not the case, as we document in our monograph using a qualitative Time Series Analysis.

We present the decisions made by the Soviet Union's Communist Party congresses and Central Committee meetings concerning the task goals of psychological research in different sub-fields of psychology (e.g. Human Experimental Psychology, Social Psychology, Personality Psychology). We relate the political conditions of a given period to the percentage of pages of psychological papers published in different areas of psychology in Voprosy Psichologii ("Questions of Psychology"), the only Russian psychology journal prior to 1977. The nature and types of the papers reflected the vicissitudes of Soviet and Russian psychology during the latter half of the 20th century when it emerged from pedagogy in 1955, as it became a sub-field of pedagogy by a decree of the Communist Party in 1937 (London, 1949; Petrovsky, 2000).

In the following sections, we give a short sketch of the Russian and Soviet Psychology schools of thought, a history of the Russian Empire, the Soviet Union, and the Russian Federation and their impact on psychology. Then we explain our research method, it's limitations and how to read the graphical representations of our results.

In the next chapter we investigate the Communist Party's role in the development of the Soviet Union's and Russian Federation's psychology from 1955 to 2000. We present our findings of the party's influence on psychology in the post-Stalin era, based on our bibliometric research using the qualitative Time Series Analysis method. Finally, we draw some conclusions and provide an outline of our future research.

This research was aided by the support of many individuals. We especially thank the following persons:

Svetlana Yu. Zhdanova, Chair of Developmental Psychology, Department of Psychology, State University, Perm, Russia, provided us with the tables of content of Russian and Soviet psychological journals, which were the data basis for our research.

Olaf Morgenroth, Professor of Psychology, Faculty of Psychology, Medical School Hamburg, Germany, who is engaged in the History of Psychology, was the mentor and coach of our research.

Jonas Knöll, Researcher, Friedrich-Löffler-Institute, Celle, Germany gave us initial support in writing the R-scripts for the statistical evaluation and representation of our data.

Lydia Lange, Max-Planck-Institute for Educational Research, Berlin, Germany supported us with her expertise in bibliometric research.

Madlen Schmaltz, Institute of Information Systems, Leuphana University, Lüneburg, Germany was our reliable office assistant.

The Team of Interlibrary Loans, Media and Information Center, Leuphana University, Lüneburg, Germany provided us with the piles of literature we needed for the interpretation of our data.

Arlene Veldkamp and Nataliia Kumorkievich did the final proofreading. Thank you so much!

There have been much more persons, who supported our work and encouraged us to go on, who are not mentioned here. We thank them also very much.

Russian And Soviet Psychology Schools

Figure 1 displays an overview of the Russian and Soviet Schools of thought in psychology modified from Ehrsam (1985). In his paper "Zur Entwicklung einer marxistischen Persönlichkeitspsychologie in der UdSSR (On the Development of a Marxist Personality Psychology in the USSR)" he identifies three main roots of Russian and Soviet Psychology. They are based on Ivan Sechenov's reflex theory, Wundt's experimental psychology and Lenin's reflection theory.

Sechenov proceeded in his theoretical and experimental investigations from the principle of materialistic monism, i.e. the inseparable connection between the psychic and the physiological, as well as from the principle of the interaction between organism and environment. On this basis, in his investigations of the reflex process going back to Descartes, he worked out the central role of information reception and discovered the inhibiting influence of the brain on reflex activity, thus referring to the possibility of active, selective behavior of the organism.

Figure 1: Russian and Soviet Schools of Thought modified from Ehrsam (1985)

Based on this, Sechenov can be considered the founder of the reflexive conception of the psychic, which had an impact far into later Soviet physiology or psychology. Sechenov published his theories in 1863 (Graham, 1972), which was banned by the Tsar's censorship. This theory was the basis for both Bekhterev's Reflexology and Pavlov's Higher Nervous Activity concept.

The second root was the German Wundt school, which was represented by Chelpanov and his disciples.

The third root was the principle of Reflection Theory, formulated by Lenin (McLeish, 1975, p. 178), following Marx and Engels. In this theory the basis is the conviction, that the mind, or psyche, is a 'reflection' of the objective, external world. According to this view, the mental and spiritual life of man is ultimately the product of social influences. The economic structure and class relations of society constitute by far the most important part. The emotions, the will, the intellect are considered to be processes which arise, develop, and change in their manifold ways as a consequence of particular material and social environments. The subjective life of a man is not something 'locked away', something personal, or unique, or individual, and developing in isolation from reality, with a special history of its own, out of contact with the real changing world of physical and social relations. Soviet psychologists cannot conceive a man as a contemplative being: human qualities presuppose a world of interaction and human relations.

In the initial stage of Soviet psychology, already at the Second All-Russian Congress of Psychoneurology in Petrograd (now St. Petersburg) in 1924, two main currents crystallized: the reactological direction led by Kornilov and the reflexological direction developed by Bekhterev and his strong scientific school. The Wundt school, represented by Chelpanov, played a minor role. In addition, at the beginning of the 1920s, Pavlov's theory of higher nervous activity gained increasing importance. All the currents or schools mentioned here were based to a greater or lesser extent on the reflexive conception of the psychic founded by Sechenov - a proof of the strong appeal of this theory and its usefulness for working out new theoretical foundations for psychological research.

One theory that gained many adherents in the 1920s was reflexology, created by Bekhterev. On the basis of Sechenov's conception of the reflex activity of the brain, Bekhterev tried to explain human behavior with the help of reflexes; in his view, only the reflex method was considered scientifically acceptable for psychology and physiology. Therefore, he limited the subject area of his reflexology to the connection of external stimulus effects and externally visible reactions and thus reduced psychology to physiology or reflexology.

With parts of his theory of higher nervous activity, especially the theory of the second signal system, Pavlov crossed the border between physiology and psychology. Thus, in contrast to Bekhterev, he did not deny the existence of psychology. Pavlov's theory could thus serve as foundation for developing psychophysiology, based on which such scientists as Bernshtein, Anochin, Merlin and others could build their discoveries.

However, due to its great contradiction between its Marxist theoretical building and experimental-methodical work, reactology could not provide basis for the new tasks assigned to psychology by social practice either. These tasks can be characterized by the completion of the socialist reconstruction of the people's economy and agriculture and the related struggle for the improvement of the material situation of the working people. Thus, there was no doubt in necessity to create a new psychological theory and that it was not enough to modify the old teachings. According to Vygotsky, the very foundation of psychology must be transformed. Thereby the Soviet psychology entered the second phase of its development - the phase of emergence of Marxist psychological theory - which lasted approximately from 1931 to 1945 (see Budilova, 1975).

The theoretical search proceeded in two directions. At the beginning of the nineteenthirties the problem of social-historical conditionality of the psyche became the main problem. It was worked on in the cultural-historical school founded by Vygotsky and his collaborators. The most important of these was A.V. Leontiev.

Kornilov, under whose leadership such psychologists as Vygotsky, Luria, Teplov, Smirnov and others gathered at the Moscow Institute of Experimental Psychology, was the first Soviet psychologist having attempted to create a comprehensive Marxist program for psychology.

As a starting point for emerging of the system of Marxist psychology, he submitted a thesis that the psychic is a property of highly organized matter, that its essence consists in the reflection of the material world. He defined physiological and psychological phenomena as two sides of a unified process, although the psychic would have an independent quality. Both, physiology and psychology had to study reactions, but the main difference between them, in Kornilov's view, was the fact that psychology had to start from the social meanings of reactions, while physiology had to abstract from the social components. In this context, he preferred the concept of reaction to that of reflex, since the reflex concept excluded any subjective content, while the reaction concept grasped the physiological-reflexive as well as the psychological processes in their unity. However, there was a great contradiction between Kornilov's general Marxist claim and the concrete results obtained by conducting experiments going back to Wundt, for measuring the reaction time, intensity and form of movement to an external stimulus (see Ehrsam, 1985).

Later, in the middle of the same decade, the problem of consciousness and activity came to the fore. A thorough analysis of the psychological problems in Marx's works and the study of his conception of activity made it possible to combine the thoughts about psychology expressed by him at different times in different works into an internally coherent system of ideas. With the help of this philosophical-deductive approach, in 1934 Rubinshtein expressed the idea of unity of consciousness and activity and psychophysical unity as two essential principles for the construction of a Marxist psychological theory and substantiated it in his book "Fundamentals of Psychology" (Ehrsam, 1985).

On the basis of these principles researches were carried out on different levels. In addition to the already mentioned psychologists like Vygotsky, A.N. Leontiev, and Rubinshtein, Teplov chose the concrete problem of abilities and deepened more and more into the experimental investigation of the individual differences between people. Ananyev created an experimental school dealing with the problems of sensory reflection, while Uznadze in his theory of set (i. e. attitude) studied the role of unconscious mental states in human activity.

Besides the schools mentioned, the psychophysiological research based on Pavlov's theory was also being advanced due to Anokhin, who dealt with the mechanism of conception of disturbed functions of the organism, which ultimately led him to the principle of reafference. Bernshtein was the most radical critic of the reflex concept.

The school deserving special attention for its theoretical level is the culture-historical school founded by L. S. Vygotsky, already mentioned above.

Vygotsky consistently proceeded from the social genesis of psychic functions and activities. He considered a man to be qualitatively different from an animal. In this framework, the stimulus-response scheme was not sufficient for Vygotsky to study the higher mental functions, since it could not capture the historical and social experience of a man. He assumed that in the course of their social development, especially through their labor activity, people artificially created new stimuli, which he called signs. Such signs were, for example, words, numbers, characters, works of art, maps, etc., and served as ways of social communication. These new, socially created artificial stimuli made self-stimulation possible and thus made a man relatively independent from the immediate external stimulus influence. This is how, according to Vygotsky, the simple stimulus-response scheme of the predominantly physiologically oriented psychologists was significantly expanded by the introduction of the sign system (Ehrsam, 1985).

Based on this theory of signs, Vygotsky developed the second pillar of his theory - the interiorization concept. He asked himself how the socially created signs are appropriated by the individual and came to the conclusion that they are gradually interiorized in the course of ontogenesis.

Vygotsky used the term interiorization to capture the process of transformation of an external action into a mental activity. Accordingly, the ideational activity is a phylo- and ontogenetically seen product of a preceding material activity. Besides his interiorization concept and the sign theory, his historical method was the most important contribution to the discussion at that time about the development of Marxist psychology. Here, in contrast to the then prevailing procedure of philosophical deduction, he attempted to analyze the essence of psychic processes with the help of historical reconstruction (Ehrsam, 1985).

After Vygotsky died, these fundamental ideas were essentially continued by his student A.N. Leontiev. The concept of appropriation played a central role in his early works. According to this, in contrast to the adaptive behavior of animals, humans actively appropriate the specifically human experience, which is of socio-historical nature, by engaging in an active confrontation with the objective world and interaction with other people. Through this process of appropriation, the individual reproduces historically formed abilities and skills.

This concept of appropriation was specified in the later works in the form of the concept of activity. With the help of the category of "representational activity," Leontiev attempted to overcome the stimulus-response scheme that had hitherto predominated in the works of Soviet psychologists by a tripartite scheme with the middle member of subject-bound activity.

In doing so, he defined activity as a unit of life mediated by psychic reflection, whose real function is to orient the subject in the representational world (Ehrsam, 1985).

Another important theory, which had an impact on Soviet psychology was formed in the 1930s and was devoted to the study of non-conscious psychic phenomena in their connection with the activity of the individual - thus an attempt to describe unconscious psychic processes, which was contrary to the Freudian theory. The theoretical head of this school was Uznadze, also a student of Wundt like Chelpanov (Ehrsam, 1985). In his theory of set (i. e. attitude), the role of nonconscious mental states in people's actions was revealed, expressing the personality's readiness for a certain activity in an emerged objective situation. The research conducted on the basis of the principle of unity of consciousness and activity made two major contributions to the developing of Marxist psychological theory: first, the elucidation of individual characteristics of unconscious information processing, and second, the description of the subject function of personality in the concept of "objectification". By this Uznadze understands the conscious human behavior towards the environment, the precondition of which is the unconscious "ustanovka" (i.e. set or attitude) mechanism (Ehrsam, 1985).

An other essential direction of psychological research in this second stage of the formation of Marxist psychology in the Soviet Union was increasingly oriented on the psychophysiological investigations of the Pavlov school. Anokhin, Bernshtein, Orbell, Kupalov and others contributed to the research. Anochin's contribution must probably be evaluated as the most significant one (Ehrsam, 1985).

Since 1930 Anochin's laboratory had been engaged in the mechanism of compensation of mental punctures of the organism, which also gave information about the principle of reafference, already discovered in 1935. According to this principle, a copy of every efference and every command from the brain to the executive organ is made, which is comparable with the reactivity when the reaction is being carried out. Thus, the brain has the possibility to control the success of the action being performed and to correct it if necessary. According to Anokhin, the reafference extended the reflex arc by a fourth link and completed it to the reflex circuit. It explains the process of feedback in terms physiology and helps clarify the question, how the organism can decide whether the sanctioning afference satisfies the criteria of adaptation to the environment.

This "cybernetic" extension of the reflex arc, long before the discovery of cybernetics, points to the seminal theory of reafference, which ultimately leads to today's "modern" theories of action regulation. (Ehrsam, 1985)

Thus, in the course of profiling of psychology in the phase lasting until the end of the Soviet Union, three centers of institutionalized psychology were formed in Moscow, Leningrad and Tbilisi. At Leningrad University, a broad experimental study of sensory reflection - sensations, perceptions and ideas - was carried out; in addition, general psychological studies were conducted under the direction of Ananyev. In the second half of the 1950s under Lomov's leadership the research in the field of industrial and engineering psychology as well as social psychology began. The Psychology Department of Moscow University, founded by Rubinshtein and later headed by A. N. Leontiev, developed the theory of the historical conditionality of the psyche of a man and the theory of the functional systems realizing the higher psychic functions. At the same time, Galperin and his colleagues developed the theory of the formation of mental actions. Luria elaborated a new branch of psychology - neuropsychology. The research work of the Chair of Psychology at the Pedagogical Institute in Moscow, which was headed by Kornilov until his death in 1957, then by Dobrynin and Petrovsky, among others, focused on the problem of personality.

The theory of set (attitude), established by Uznadze, was further being developed in Tiflis (Georgia) by Prangishvili and Natadze (Budilova, 1975).

History Of Russia And Sociopolitical Influences On Psychology

To provide a better understanding of the social and political environment during the different eras of Russia (until 1917), the Soviet Union (1917-1991) and the Russian Federation (1991-today) investigated in our research, we offer short sketches of the relevant political leaders, namely, Tsars, leaders of the Communist Party, and Presidents of the Russian Federation.

This description can only be sketchy, as we have to condense the many persons to those, whom we consider essential in order to grasp the essence of what happened over time. If one has the desire to dive more deeply into Russian and Soviet history, we recommend reading the book "A Concise History of Russia" (Bushkovitch, 2012).

As the historian W. Taubman said "Three issues - relating to political labels, records of meetings of the ruling Communist party Politburo, and transliteration of Russian language - deserve special attention" in the historical research on Soviet Union (Taubman, 2017, p. XI). This statement of W. Taubman reveals the problems every researcher is confronted with while doing historical research on topics related to Soviet times in Russia. It is also true for our research on Soviet and Russian psychology. If one delves into the literature dealing with this area, one finds three different versions of the development and application of psychology in Russia and the Soviet Union - depending on the sources on information one uses. The first source is the literature written in Soviet times by authors raised in Soviet Union and stayed there (e.g. Davydov, 1982; Rubinshtein, 1971). In this source Russian and Soviet Psychology was narrated as a continuing development directed by the Communist party of the Soviet Union. The second source is the literature written by western authors raised in the western countries (mainly in the US, UK, and Germany) in the times of the Cold War after the

Second World War (e.g. Bauer, 1959; Graham, 1972; Kussmann, 1974; McLeish, 1975; Thielen, 1984). Most of them spent some time in the USSR and had access to local information available during their research. They describe the history of Russian and Soviet Psychology as a chain of trials and errors, of promotion and repression of psychologists and psychology as a whole, and of many sudden turning points in Soviet Union's development over times. The third source of literature is from authors who were raised in the USSR and had access to the archives of the Communist Party and the administrational bodies of the Soviet Union from the years 1986 up to now (e.g. Bratus, 1998; Kozulin, 1984; Krementsov, 1997; Petrovsky, 2000). The authors of this third group describe the development of psychological science in a way similar to how the second group did it. However, the ones who remained living in Russia (Bratus and Petrovsky) drew a softer image relative to that of the Western authors of the political influences of the Communist Party, the fate of psychologists and the psychological science as a whole in the changing political environment. This is mainly true for the first repressions against psychologists (1929-1931) and the next waves of repressions against psychology in Stalinist times (1936 – 1953). As the narrative of the first group differed in many respects from the ones of the latter two groups, we relied mainly on the reports presented by the latter two groups. We did this assuming that the censorship of the Communist Party forced the alteration of historical facts, as was the case of the events during the "October Revolution" in 1917 and of Stalin's role in the revolutionary process in Russia and the Soviet Union.

In order to understand this problem better, we take the report of psychotherapy by the two sides as an example. How different was it described by Soviet and East-German authors versus by their US and West-German counterparts? This was how East-German psychiatrists said of the Soviet care of the psychologically sick people: "Psychiatric care in the USSR has been developed to a certain perfection for a very long time" (Eichhorn & Stern, 1977, p. 578). And (Lauterbach, 1976, p. 225) concluded "Clinical psychologists in the

Soviet Union call themselves 'Pathopsychologists'. They mainly deal with diagnostic questions and prepare expert opinions for various purposes. ... In contrast to our clinical psychologists, however, they are not involved in psychotherapeutic tasks" (Eichhorn & Stern, 1977; Lauterbach, 1976).

After the opening of the archives of the Soviet Union following the "Glasnost (i.e. openness)" efforts of Gorbachev, it was revealed that the Western view (the latter two groups mentioned above) of the situation was closer to the truth compared to the Eastern views (first group of authors from above). This is confirmed by our bibliometric research in the following sections of this chapter.

Tsarist Times

Political Environment

The Russian empire was founded in 1547 by Ivan the Terrible (Massie, 2013), who called himself "Grand Prince of Moscow". His descendants eventually called themselves "Tsar", which is the Russian word meaning "Emperor" in English, or "Kaiser" in German, or "Caesar" in Latin. The Tsar was the absolute ruler of his country and Moscow was considered the "Third Rome" after Constantinople (now Istanbul in Turkey) in the East Roman Empire and Rome in Italy, the capital of the Roman Empire (later the West Roman Empire). Rome is also the capital of the Roman Catholic Church. Constantinople was the capital of the Greek-Orthodox Church, and Moscow was and is now the capital of the Russian Orthodox Church.

The Russian Empire stretched from the Baltic Sea to the Pacific Ocean, covering 1/6th of the Earth's land. Innovations were brought into the country by Tsars, such as Peter I (1672 - 1725), who wanted Russia to develop into a European State. He founded the new capital, St. Petersburg, which was called the "Window to Europe" or "Venice of the North", because of its large number of channels, which were built to drain the swamp area where the capital was founded. This new capital at the Baltic Sea was designed by European architects and has a port for his navy and for trade with European countries.

The last Tsar was Nicholas II (1868 - 1918), who was inaugurated in 1894 after the death of his father Alexander III. His grandfather Alexander II (1866-1881) had started some political reforms, after several revolts, beginning with the "Decabrist" uprising of army officers in 1825. Alexander II was murdered by a bomb blast and his son Alexander III tried to reverse the political reforms of his father, which installed more political participation by the people. Like his predecessors and the Emperors of Austria and Germany Nicholas II was convinced that he was put in his position by God and God will guide to him to make the right decisions (Dickinger, 2001; Hamann, 2010; Massie, 2013; von Rauch & Geierhos, 1990). According to their opinion peoples participation in politics was not considered by God.

Figure 2:
Tsar Nicholas II
Public Domain,
https://commons.wikimedia.org

In the beginning of the 20th century, the majority of the Russian population was still former rural serves, who had received no education at all. More than 90 % of the population was living in rural areas, where the Mir (village-council) supervised them and their activities. According to McLeish (1975), the Tsar's secretary of enlightenment said that knowledge was of worth only if it was applied like salt. In other words, it should be given only in small amounts according to the urgent needs of the population. For the majority of the people, education would bring more harm than benefit.

This was the background in the Russian empire, when Nicholas' II succeeded to the throne. The middle class (bourgeoisie) continued to be dissatisfied with the Tsar and his power. There was a mass demonstration in St. Petersburg in front of the Tsar's palace in 1905, over the severe food shortage caused by the Russia's war with Japan, in which Russia was defeated. This mass demonstration was dissolved by the army, shooting and killing many demonstrators. This "bloody Sunday" was the start of several revolutionary uprisings, which was followed by the "February" Revolution in March 1917 with the Tsar's abdication and culminated in the Bolshevik Communist so-called "October Revolution" on November 7th 1917 (at that time the old Julian Calendar was used in Russia, which was behind the now Gregorian Calendar: Gregorian November 7th was October 25th in Julian Calendar). For more details about October Revolution see Altrichter (2013), Khlevniuk (2015), Lewin (2016), Sebestyen (2017), Tucker (1974).

What were the Conditions like for Psychology in this Era?

Russian psychology and physiology, which were closely connected to philosophy, evolved in the second half of the 19th century. At these times life in cities was under the tight supervision of the tsarist secret service and consequently all publications had to pass the state censorship.

Figure 3:
Georgy I. Chelpanov
Public Domain
https://commons.wikimedia.org

The Tsar's censorship was not only applied to political publications of every kind, but also to science and all other research. During this era psychologists and physiologists like Chelpanov, Sechenov, Bekhterev, and Pavlov published their first works in physiology, which adressed in part also the emerging subject of psychology. Sechenov wrote in 1863 "Reflexes of the Brain". There he expressed his opinion of

Figure 4:
Ivan M. Sechenov
Public Domain
https://commons.wikimedia.org

the unity of the body and the soul, which was not in accord with the Orthodox Church's ideology of an immortal soul and a mortal body. Therefore the publication was halted by the Tsarist censorship, which was also under control of the Orthodox church (Kussmann, 1974).

Figure 5:
Vladimir M. Bekhterev
Public Domain
https://commons.wikimedia.org

During this time, many Russians had close connections to European scientists, and some of them studied at the laboratories of Wundt and Fechner in Germany, and graduated from there. One example was Bekhterev, who attended the 3rd international psychological conference in Munich, where he had intensive discussions with his German colleague Theodor Lipps (1851-1914) (Fritsche, 1980).

Figure 6:
Ivan P. Pavlov
Public Domain
https://commons.wikimedia.org

Because of the repression on scientific and psychological research most of the researchers and university staff were hoping for more freedom through replacing the Tsar's rule by democracy. So, they welcomed the Tsar's abdication in March 1917 (the so-called February Revolution) and hoped that freedom would emerge after the Bolshevik revolution (the so-called October Revolution) in November 1917 (Bauer, 1959; Bratus, 1998).

From Bolshevik October Revolution In 1917 To 1929

Political Environment

According to Sebestyen (2017), the Bolshevik Communist Party's victory in the October Revolution (see above) was a surprise. The victory was not based on Lenin's plans but rather due to the weakness of the bourgeois (i.e. middle-class) government, which was installed by the Duma (Russian parliament) after the Tsar's abdication in February 1917. The Communists had no plan to rule and the only established structures were those of the party. The main ruling body was the Central Committee, which had 6 members (including Stalin) and was informally headed by Vladimir I. Lenin (1870-1924). A second committee was the 7 member Politbureau, whose task was to co-ordinate the political actions of the Party. Besides Lenin, Joseph V. Stalin (1878-1953), and Leo D. Trotsky (1879-1940) were two leading revolutionists. And there were the Party Conventions, which were party's general assembly (Lewin, 2016; von Rauch & Geierhos, 1990).

Figure 7: Vladimir I. Lenin
Public Domain https://commons.wikimedia.org

Lenin was born as Vladimir I. Uljanov, the son of a high school teacher, who was awarded the title of a hereditary count for his efforts. He became a revolutionist after the execution of his elder brother Alexander, who was involved in an assassination attempt at Tsar (Sebestyen, 2017; Shub, 1962). He spent much of his lifetime in Siberia (the vast eastern part of the Russian empire), in Finland, Switzerland, and Germany. He was a believer in the theories of Marx and Engels. His life was dedicated to the world-wide socialist revolution, which was expected to happen in the industrial countries such as England and Germany.

In April 1917 he returned to Russia's capital St. Petersburg in order to promote the socialist revolution, despite the fact that Russia was not industrialized at that time like other European countries. He took advantage of the dissatisfaction of the majority of the population with Russia's involvement in WW I (Lewin, 2016; Sebestyen, 2017; von Rauch & Geierhos, 1990).

At that time the ruling body of Russia became the Soviets, i.e. councils of industrial workers, soldiers and rural workers. They elected the government, called "Council of Peoples Commissioners". Lenin became president of this council, Trotsky Secretary of State, and Stalin Secretary of the Russian Nationalities (Russia was and is a multinational country). There was also a Central Executive Committee of Soviets, which represented the legislative branch, the party congresses. Its head was also the President of the country.

Figure 8:
Leon D. Trotsky
Public Domain
https://commons.wikimedia.org

Soon the new government started to abolish the private properties of lands and distributed the farmland to the rural workers. Private production and trade were prohibited. Eventually, Russia ended the involvement in WW I, and signed a treaty which gave lands to previous enemy countries like Poland and Germany.

The communist party established censorship, tighter than before, to suppress the counter revolutionary ideas. The new secret police had the permission to do everything for securing Communist power including assassination and torture of real or assumed political enemies. The targeted people included every citizen of the country, believed to be not in favor of the Communist government (Altrichter, 2013; Sebestyen, 2017; von Rauch & Geierhos, 1990). In June 1918 the Tsar, his wife, their 5 children, and their servants were executed and their corpses thrown into an old mine shaft.

Leo D. Trotsky (1879-1940) established the "red army", who's aim was eliminating all enemies of the Communists in the country. There was a three-year communists-versus-royalists civil war. This civil war was won in 1921 by the Communists by means of employing terror and cruelty. The adversaries were royalists, the white guards (which were supported by the US and European countries), and members of the other political parties (Sebestyen, 2017; von Rauch & Geierhos, 1990).

The civil war left whole country in chaos. About 5 million people died due to famine. Orphans assembled into criminal gangs, and attending school was elective. In every aspect of public life was a lack of discipline.

In order to improve economy and to supply the population with goods, Lenin installed the New Economic Policy (NEP), which permitted private production and trade in smaller scales again. This improved the living conditions of the population.

Another effort was put into the alphabetization of the population, which was a very slow process. The main goal of this was the indoctrination of the people with the Bolshevik ideology and giving them the Bolshevik view of history.

Lenin had his first stroke in 1922. He was paralyzed on one side of his body and died in January 1924. His body was embalmed and exhibited in a new mausoleum at the Kremlin's wall (Kremlin is the old fortress of Moscow).

No successor was appointed by Lenin and the country (now called Soviet Union) was ruled by the Central Committee of the Communist Party (with Stalin as the First Secretary) and the Council of People's Commissioners (government). Gradually Stalin achieved more power and in 1929 he became the de facto head of the Communist Party and the dictator of the country (Khlevniuk, 2015; Tucker, 1974; von Rauch & Geierhos, 1990). In 1928, he presented the first 5-year plan for the industrial and agricultural development of the country.

What were the Conditions like for Psychology in this Era?

The Communists believed in the power of science and supported it by securing the scientists not only the resources for research but also food, housing, and other materials. New schools of thoughts in psychology were established, new psychological journals started to be published, Soviet scientists became active in international conferences, and many new related publications were also printed (Bratus, 1998; Petrovsky, 2000). Ironically, although the newspaper, journal and book censorship employed by the Communist party was much stronger than at Tsarist times, it didn't apply to scientific publications. The psychologists and physiologists from the Tsar's era enjoyed their freedom. According to Bratus (1998), the number of published psychological books in Soviet Union peaked in 1929 at 600.

Figure 9:
Konstantin N. Kornilov
Public Domain
https://commons.wikimedia.org

Figure 10:
Lev S. Vygotsky
Public Domain
https://commons.wikimedia.org

During this period, Kornilov established his Reactology school, and Vygotsky, Luria and Leontiev established the cultural historical school of psychology (Bauer, 1959). In addition, there were more practical efforts to test people for the purpose of assigning them to the suited school career in children's education (pedology), or the best suited persons to the open positions in army, administration, academia, and industry (psychotechnics). From the beginning of the Communist rule, one lofty goal for psychology was the creation of the new type of men and their fairer assignment to positions according to their abilities, not according to their social class as was the case in the Tsarist times (Davydov, 1982).

According to Hyman (2017) there was an international attitude in the Soviet Union in the 1920s and early 1930s. As she said, "The ideological and intellectual climate in the aftermath of the Revolution was cosmopolitan" (p. 639). At that time, the Soviet psychologists had tight connections to their Western colleagues and were fluent in foreign languages, e.g. „Vygotsky's writings were densely filled with references to Western psychologists (such as Adler, Bühler, Claparède, Freud, James, Janet, Köhler, Koffka, Lewin, Piaget, Stern and Werner)" (p. 632).

Figure 11:
Alexander R. Luria
Public Domain
https://commons.wikimedia.org

There were some cases of repression until 1929. Many scholars emigrated to Western countries such as UK, US, France, and Germany. Those who did not emigrate or were assumed to be not in favor of the new rulers were imprisoned or executed.

Figure 12:
Alexander N. Leontiev
Public Domain
https://commons.wikimedia.org

One example is Chelpanov, the director of the Moscow Institute of Psychology was replaced by his disciple Kornilov (Kozulin, 1984), who claimed that Chelpanov's theories were not based on dialectical Materialism. Another case was the murder of Bekhterev, who was found killed by poison in his Moscow hotel two days after he met with Stalin and diagnosed Stalin with a severe Paranoia in 1927 (Kesselring, 2011).

First Wave Of Repressions Against Psychologists In 1929

Political Environment

Jospeh V. Stalin (1878-1953) was born in a poor family as Joseph V. Jugashvili in Gori, a small village in the Caucasus mountains of Georgia. His father was often drunk, violent and absent from home. He followed his mother's desire for him to become a priest of the Russian Orthodox church. In the environment of the clerical school he became acquainted with revolutionists and from then on became active in the revolutionary underground (Altrichter, 2013; Khlevniuk, 2015; Lewin, 2016; Tucker, 1974; Zubok, 2009). There he started to call himself "Stalin", which was intended to be seen hard and enduring like steel.

Figure 13: Joseph V. Stalin
Public Domain https://commons.wikimedia.org

Several times he was imprisoned and sent into an exile in Siberia, but other than many revolutionaries, he stayed mainly in the Russian empire. He supported Lenin, and in an opportunistic manner, was looking for his opportunities to be promoted within the Bolshevik system. As Zubok (2009) said, Stalin was "Always an opportunist of power, he succeeded at home by allying with some of his rivals against others and then destroying them all (Position 830)".

This was the means by which he attained, step by step starting in 1929, the dictator's power in the Soviet Union. From 1929 to 1932 he initiated political purges against everyone in the Communist Party, that was not favoring his power or that learned about his weaknesses.

He reversed the NEP (New Economic Policy), which Lenin had started to remedy the economical shortcomings of the centralized planned economy. Those, who had profited from doing businesses, were now branded as enemies of the people and sent to forced labor camps, or executed. He also ordered the farmers to form government-controlled co-operatives. This led to new famine waves claiming the lives of about another 10 million victims. He fought the famine by forcing the farmers to carry out excessive grain deliveries. Those who resisted were either sent to forced labor camps or executed. Ironically, at that time, the Soviet Union was a big grain exporter, as the country would have had no money for purchasing the machinery needed for the newly established industrial plants (von Rauch & Geierhos, 1990).

Using the workforce of the forced labor camps, Stalin started to industrialize the country with electrical power plants, steel factories, new canals for cargo ships, railways, and new mines of coal and metals (Khlevniuk, 2015; Tucker, 1990; Zubok, 2009).

What were the Conditions like for Psychology in this Era?

The first wave of political purges affected also many psychologists. The main goal of the purges was "to defeat the remains of bourgeois theories and to destroy them, which reflect directly the counterrevolutionary elements against the socialistic installation of the country" (Bratus, 1998, p. 6). One of the victims was Vygotsky, who was accused of harboring elements of "bourgeois influence" and "perversion of Marxism" in his theories (McLeish, 1975, p. 121). He was removed from his position at the Moscow Institute of Psychology. In 1931 there was a discussion of Kornilov's "Reactology", in which Kornilov was accused of using "mechanistic concepts", committing "severe ideological errors" and making "compromises between subjectivism and objectivism". As a result, Kornilov was dismissed as director of the Moscow Institute of Psychology (Krementsov, 1997, p. 27). Similarly, some schools of psychotherapy for example, psychoanalytical therapy, were banned as "idealistic" and "subjectivistic" (These were created to have a reason for prosecution).

Figure 14:
Sabina N. Shpilrein
Public Domain
https://commons.wikimedia.org

One of the victims has been Sabina Spielrein (1885-1942). She was a medical doctor, who received her degree at the University of Zurich (Switzerland) and became one of the leading psychotherapists and psychoanalysts in Switzerland, Austria, and Germany. In 1923 she returned to Russia and worked as a psychoanalyst in her home city Rostov on the river Don (Covington & Wharton, 2015; Richebaecher, 2008). In 1929, psychoanalysis was prohibited, as explained above, and she started to work in Pedology, i.e. mainly testing children. However, as described in the next paragraph, in 1937 Pedology was banned in the Soviet Union, and she switched to work as a medical doctor. In 1942 during WW II, Rostov on Don was conquered by German troops and Sabina Spielrein, whos parents were Jewish, were executed together with 27,000 other mostly Jewish victims (Covington & Wharton, 2015; Richebaecher, 2008).

There were many more psychologist victims. However, it is not easy to distinguish persecutions specifically for psychological ideas and thoughts from general political purges which often led to victims sent to forced labor camps, prisons, or even death penalty.

Despite the repression, a highlight of the international co-operation in psychology was the first international psychological conference held in the Soviet Union in September 1931 in Moscow. The theme of this conference was "Psychotechnic", a subject which was later banned in 1936 (Volkov et al., 1988) as described in the next paragraph.

Second Wave Of Repressions Against Psychology In 1936

Political Environment

From 1929 to 1936 Stalin had stabilized his power and started to eliminate those, who could rival him or had knowledge of his weak points. He started to improve relationships with foreign countries which culminated in the Treaty of Non-Aggression with Nazi-Germany. Now he started his purges, which affected also psychologists.

What were the Conditions like for Psychology in this Era?

In July 1936, a decree from the Central Committee of the CPSU (Communist Party of the Soviet Union) condemned those educational psychologists who had been engaged in pedological studies and testing. Zalkind and Vygotsky (who died in 1934), as well as Kolbanovsky and Blonsky, were put on a blacklist (Petrovsky, 2000). Given that almost all work in educational psychology in the 1920s was called "pedology", one may imagine the consequences of this decree. This decree affected also the so-called "psychotecnic", which was testing professionals. All forms of intelligence testing and other applied studies fell victim to the witch-hunt and were subsequently forbidden (Kozulin, 1984). According to Petrovsky

Figure 15:
Aron B. Zalkind
Public Domain
https://commons.wikimedia.org

Figure 16:
Viktor N. Kolbanovsky
Public Domain
https://commons.wikimedia.org

(2000, p. 22) psychology was now "castrated". In the university textbooks during these years the authors tried to prevent future teachers from learning anything related to "child", "pedagogical", and "school" psychology. The Russian psychologists got only reduced knowledge about psychology during this period. As a result, in 1936 all laboratories for pedology, psychotecnic and industrial psychology were closed down and all the work in these areas ceased.

The application of the so-called pedology-decree was not limited to the areas mentioned above (Keiler, 1988). It was also applied to psychological areas not connected with pedology. It caused the complete dissolving of the institutionalized psychology: the Soviet Society of Psychologists was closed down, all psychology journals stopped publication, there were no conferences on psychology anymore, and finally no public discussions on psychological matters were permitted.

Figure 17:
Pavel P. Blonsky
Public Domain
https://commons.wikimedia.org

Thus, the new kind of Soviet psychology was the main winner of the fight in 1936 (Yasnitsky, 2016). From now on, Psychology was a part of pedagogy with the main goal of creation of the new man. This was followed by a series of other important achievements, that included the publication of a range of officially endorsed textbooks in psychology in 1938-41, the establishment of the Institutes of Psychology in Soviet Geor-

Figure 18:
Sergej L. Rubinshtein
Public Domain
https://commons.wikimedia.org

gia (under the auspices of the Academy of Sciences of GSSR, in 1941) and in Soviet Ukraine (in 1945), the granting of important and the most prestigious national scientific awards to psychologists (e.g. the award of the Stalin Prize to S. L. Rubinshtein in 1941), and the first appointments of psychologists to the top of the social scientific hierarchy, the Academy of Sciences of the USSR (Rubinshtein and Kravkov as "Corresponding Members") in the 1940s. Finally, the culminating event arrived: In 1946 the "castrated" Soviet psychology was introduced in public school curricula as a mandatory subject to be taught all over the Soviet Union. This event logically concluded the ten-year period that can be legitimately referred as the Golden Age of Soviet Psychology from 1936 to 1946 (Petrovsky, 2000).

Pavlovization Of Psychology From 1948 To 1953

Figure 19:
Trofim D. Lysenko
Public Domain
https://commons.wikimedia.org

But new trouble for psychology was not far away. At the end of July 1948, VASKhNIL (Всесоюзная Академия Сельско Хозяйственных Наук имени В. И. Ленина, in English translation: V. I. Lenin All-Union Academy of Agricultural Sciences) held a meeting "On the Situation in Biological Science" (Artemyeva, 2015; Joravsky, 1986; Petrovsky, 2000). The main Speaker was Trofim Denisovich Lysenko, the head of the academy. It marked the beginning of a new setback not only for biological sciences, but for science as a whole (Krementsov, 1997). From now on, a clear distinction was made between the "bourgeois western science" and the "progressive socialistic science". It started with that meeting, where the genetic theories of Mendel and Morgan were replaced by the theories of Lamarck and Michurin, who, unlike Mendel and Morgan, believed in inheritability of acquired traits. Darwin's theory of natural selection was branded as bourgeois (i.e. non-communist, middle-class, capitalistic) as well. This meeting was followed by a series of other meetings, held by all scientific, educational, and medical institutions throughout the country. "Stalin's sentence uttered in 1948 at the Politburo sitting in June – 'The Central Committee can have its own position on scientific questions' - signified a serious change in the posture of the party leadership toward science and the scientific community; the scientific community would no longer be granted authority and autonomy in scientific matters" (Krementsov, 1997, p. 182).

In February 1949, Pravda published an article "About One Unpatriotic Group of Theater Critics," which opened the campaign against "cosmopolitanism" (Krementsov, 1997). From then on only references made to Russian - or at least Soviet - authors and papers were permitted in scientific papers. After the 100th birthday celebrations for Ivan Pavlov in September 1949 came the next stage of repression of psychologists. From that time on, psychology had to be based on "conditioned reflexes". For example, thinking had to be explained as "higher nervous activity", one of Pavlov's ideas, and speech was now the "second signaling system" which could be used as a stimulus for "conditioned reflexes" (Rueting, 2002; Todes, 2015; Tucker, 1990).

In June 1950 Pravda published Stalin's article on "Marxism and Questions of Linguistics", in which he stated that Russian culture and Russian language were superior to all other languages of the world. In order to "pavlovize" psychology completely, the Academy of Pedagogical Sciences held a meeting in March 1952 "on the situation in psychology and its reorganization on the basis of I. P. Pavlov's doctrine" (Krementsov, 1997) to promote Pavlovian-based psychology.

Figure 20:
Leoinid V. Zankow
Public Domain
https://commons.wikimedia.org

One of the victims of these repressions was Sergej L. Rubinshtein (1889-1969). "In 1949 the journal Soviet Pedagogics - in an editorial, 'Raising High the Banner of Soviet Patriotism in Education', and a paper 'To Purge Soviet Psychology of Nationless Cosmopolitanism' by P. Plotnikov - accused Rubinshtein of 'worshiping bourgeois (i.e. non-communist) science' and 'insulting Russian and Soviet psychology'" (Kozulin, 1984). As in other cases, the accusations were made up, because of his Jewish ancestry.

In the next issue of Soviet Pedagogics Leonid Zankov claimed that Rubinshtein had deliberately suppressed studies by Russian authors and advocated the decadent views of bourgeois psychology. Zankov maintained that there was no need for critical reviews of such authors as Piaget, for "it is well-known that the 'theory' of Piaget is a militant attempt to depict child intelligence in an absolutely distorted form" (Kozulin, 1984). As a result of this campaign Rubinshtein lost all his administrative positions, and continued only as a research fellow at the Institute of Philosophy. Another victim was Alexander Luria, who was removed from all of his positions because of "Anti-Pavlovism" (Krementsov, 1997).

Khrushchev Era From 1953 To 1966

Figure 21:
Nikita S. Khrushchev
Public Domain
https://commons.wikimedia.org

Khrushchev (1894–1971) succeeded Stalin as First Secretary of the Communist Party after Stalin's death in March 1953. He did not do much to change the atmosphere right away, but over time the political environment gradually improved. Psychology and science were not on top of his agenda. Pressure on psychologists was relieved step by step: A new psychological journal, Voprosy Psychologii, was launched in 1955, with six issues per year. The first issues had approximately 130 pages and were published by the Academy of Pedagogical Sciences (psychology was still considered a part of pedagogy). In addition, the "Society of Soviet Psychologists" was re-founded in December 1956 (Volkov et al., 1988). It was originally founded in 1885 as "Moscow

Psychological Society," but the society ended its activity in 1922 (Poole, 2002).

Khrushchev had completed no more than four years of elementary school (perhaps only two years); over the following years, he studied at several engineering schools. He never graduated because of his political activities in these schools (Taubman, 2003). According to Zubok, he was "strikingly under-educated and erratic" (Zubok, 2009, p. 167) and had no interest in academic matters at all. His only requirement for psychologists was to build stronger links between psychological theories and the industrial and agricultural practical work in schools and higher education. Khrushchev's approach mainly affected educational psychology and human factors engineering. His era also saw a revival of statistics and research methods.

He was open to scientific co-operation with the Western countries, and from 1956 on Soviet psychologists resumed attending international conferences. According to Hyman (2017), there rose an interest in the Western psychologists in the works of the cultural-historical writings of Vygotsky, Luria, and Leontiev (see above). Their publications were translated into English and later became subjects of Western psychological research.

Brezhnev Era 1966 To 1985

In 1966, Leonid I. Brezhnev (1906–1982) became the First Secretary of the Communist Party of the Soviet Union. Compared to his predecessor, he was relatively well educated (Schattenberg, 2017). He attended not only elementary school but also Gymnasium (a German-style university-preparatory high school), where he received a tuition and fee waiver. From 1923 to 1927, he was a student at a technical college, where he graduated as an engineer (Schattenberg, 2017).

Figure 22:
Leonid I. Brezhnev
Creative Commons https://creativecommons.org/licenses/by-sa/3.0/de/legalcode

In Brezhnev's era, world politics continued between political thaw and the Cold War. His main aims were to consolidate the Soviet Union and to improve living and housing standards. In this respect, he continued the work of his predecessor, although he was much more moderate and did not come up with new ideas. The Brezhnev era was also characterized by networks of power juggling and bribery. The last years of his era were characterized by stagnancy and a slackening of party control over the sciences. As in the Khrushchev era, psychologists did not experience repression, although the control by the Communist Party and state administrators over academia was generally still very tight. Positions in psychology were filled mainly according to party-line loyalty and not according to professional qualification. As in the Stalin era, there was a tight connection between the holders of an academic position and their sponsors in the Communist Party and in public administration (Krementsov, 1997).

In the time he was in power, international co-operation of psychologists was intensified, starting with the XVIIIth international psychological conference, held in August 1966 in Moscow (Volkov et al., 1988). There were many more international conferences convened in other areas at the Soviet universities.

After his death in 1982, Brezhnev was succeeded by Andropov, who died in 1984, followed by one year of Chernenko, who died in 1985. They did not change anything.

Gorbachev Era From 1985 To 1991

In March 1985, Mikhail Gorbachev (born in 1931) became the new First Secretary of the Communist Party and thereby the new leader of the Soviet Union. He was a descendent of a peasant family, who benefitted from farm collectivization, and he was well educated. He attended the local high school and graduated there. Because of his and his father's efforts in harvesting in his home province in the north Caucasus in 1948, he had been awarded with the "Red Banner of Labor" order and his father with the "Lenin Order," which helped him in his future career. He studied law at one of the most recognized universities of the USSR, the Moscow Lomonosov State University. Over the years, he attained several positions in the Communist Party organization, and finally became a member of its governing Politburo (Taubman, 2017).

Soon after he took office, it became clear that he was a reformer, introducing the buzzwords perestroika ("reconstruction") and glasnost ("openness") in an effort to overcome the rigid and inefficient structures within the party and the state administration. This also affected research and science in a positive way. The dependence of academic research and teaching from politics weakened fast, and international cooperation between scientists increased considerably, and became common from then on.

Figure 23:
Mikhail S. Gorbachev
Creative Commons https://creativecommons.org/licenses/by-sa/3.0/de/legalcode

Although Gorbachev wanted his reforms to save the Soviet system, he unintendently helped dissolve the Soviet Union.

Yeltsin Era From 1991 To 2000

After Gorbachev's resignation in late 1991, the main leader was Russian President Boris Yeltsin (1931–2007). His ancestors were independent farmers, self-employed blacksmiths and carpenters, and they had suffered from the so-called "dekulakization," that is, the persecution of independent farmers and craftsmen (Colton, 2008).

Figure 24:
Boris N. Yeltsin
Creative Commons
https://creativecommons.org/licenses/by-sa/3.0/de/legalcode

As a well-educated civil engineer, he was promoted within the Communist Party organization for his extraordinarily successful efforts in housing construction in the Sverdlovsk (now Yekaterinburg) district and later in Moscow during the Gorbachev period.

As president of the Russian Federation he tried in December 1991 to keep the Russian leadership in the territory of the Soviet Union by founding the Commonwealth of Independent States, which never achieved the power of the former Soviet Union.

He tried to change the socialistic economic system into a market economy (Taubman, 2017).This change had disastrous consequences for the economy and the population. High inflation and unemployment rates, and the breakdown of many governmental structures followed. Science and research suffered from the financial problems. Now the scholars were free to chose their research subject and teach according their own opinion: there was now real freedom for science.

Putin Era Since 2000

In January 2000, Yeltsin was succeeded by Vladimir Putin (born in 1952), who consolidated governmental power and brought an economic recovery for the country (Zubok, 2009). He was the third son of an industrial workers couple who lost their first two sons in infancy (Myers, 2015). His parents were strong believers in communism, as were his grandparents. His paternal grandfather was the personal cook of Lenin's widow N. K. Krupskaya.

He realized his juvenile dreams of becoming a member of the secret service: first in the KGB (in Russian: Комитет Государственной Безопасности) of the Soviet Union and then in the Russian secret service FSB (in Russian: Федеральная служба безопасности Российской Федерации). When jobs in the FSB were cut, he first became a member of staff to the President of St. Petersburg University, who later became the Lord Major of this city. Putin then became his important secretary of foreign affairs and later deputy Lord Major of St. Petersburg. Eventually, in 1996, he was appointed to a governmental position in Moscow and in August 1999 became the prime minister of the Russian Federation. When Yeltsin resigned on New Year's Eve of 2000, Putin became the interim President of the Russian Federation according to the constitution. In March 2000, he was elected President by an overwhelming majority of votes.

Figure 25:
Vladimir V. Putin
Creative Commons https://creativecommons.org/licenses/by-sa/3.0/de/legalcode

Along with economic recovery, the sciences, including psychology, had improved conditions for research and teaching, too. Every major university now has a faculty or department of psychology. There is freedom of research and teaching, as in Western countries. However, the resources are very small compared to Western research institutions and universities—a fact that Western observers quickly realize when they visit their colleagues at Russian institutions.

Research Method

In this chapter we present our research approach. We outline the motivation for our research, give a short introduction into Time Series Analysis and clarify some caveats of our results. Then we explain the meaning of the data and symbols in the graphs of the next chapter.

Motivation Of Our Research

Until Stalin's death, political influence on science and especially psychology in the Soviet Union was obvious as we explained above. This changed after his death in 1953, when Khrushchev seized power. Decisions about the direction of psychological research were now discussed in meetings of the Communist Party and in decrees of the Communist Party's Central Committee. From the 20th party convention in 1956 to the 27th convention in 1986, all but one (the 22nd) party convention made decisions concerning the tasks of psychology in the Soviet Union (Table 1). The Central Committee also issued decrees concerning the tasks for psychology in November 1958 and June 1963 (see also Table 1).

These decisions and decrees were published as articles in the journal Voprosy Psychologii (Questions of Psychology), emphasizing different aspects of the role of psychology. We cross-checked the articles published in this journal with the Soviet Academy of Sciences' "Chronicle of Science and Technology" (Volkov et al., 1988) and the "Resolutions and Decisions of the Communist Party of the Soviet Union" (Hodnett, 1974; Schwartz, 1982), making sure that no decision or decree was missing. In the following chapter, we examine these decisions and their impact on the publications in Voprosy Psychologii.

To find the Communist Party's decisions as these relate to the tasks of psychology, we translated those articles in Voprosy Psychologii, which report Communist Party decisions, from Russian into English and also classified their contents according to the APA content classification scheme (Knoell & Jou, 2018; Tuleya, 2007).

To find reflections of these party decisions in the articles published in Voprosy Psychologii, we translated the titles of all articles from 1955 to 2000 (total of 7,049 articles) into English and put them in a table, with articles classified into categories according to the

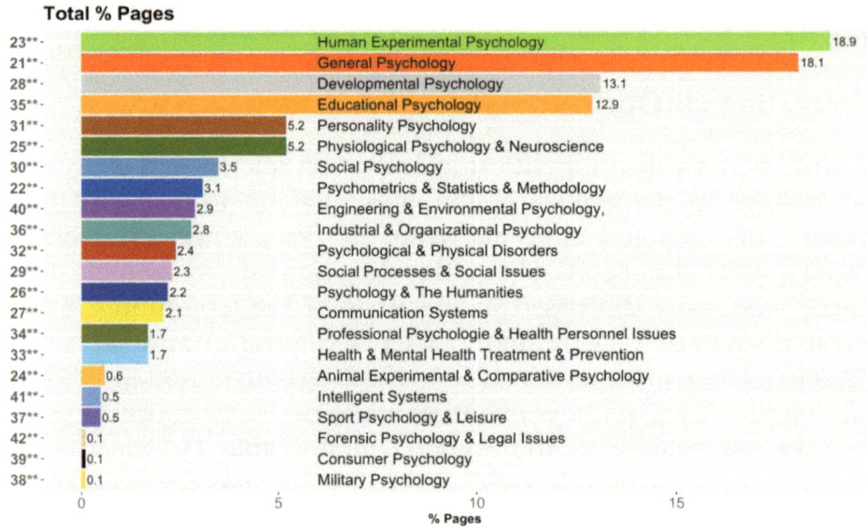

Figure 26: Percentages distribution of published content categories in Voprosy Psichologii

APA content classification scheme (Tuleya, 2007). We also added the number of pages for each article, resulting in a table featuring all article titles. This table served as the database for the R-scripts we used to calculate the statistics and draw the graphs (Field et al., 2012). The frequencies of the content categories in the period from 1955 to 2000 are shown in Figure 26.

We used the APA content classification because this classification scheme is often used for bibliometric studies. Its main advantage is its cross-cultural applicability, which makes it feasible to compare bibliometric results of multiple states or countries, such as the US and Germany (Perrez & Krampen, 2015).

Time Series Analysis

Figure 27 displays the percentage of publication in the APA content class "Personality Psychology" with the class number 31**. The two asteriscs at the end of the class number indicate, that the subclasses are also covered.

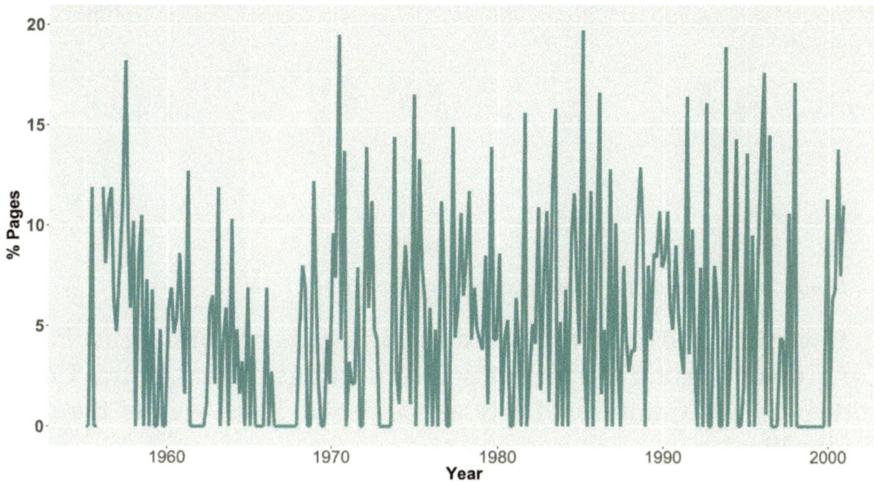

Figure 27: Percentage of Article Pages in Personalyty Psychology from 1955 to 2000

This figure shows a great variation of the percentage and there seems no trend to recognize. It is looking like a white noise over time.

In such a case a time series analysis, which is used to show an effect of a trigger or stimulus on the data of interest, is useful. So, we calculated a one-year moving average (Shumway & Stoffer, 2017) using R-scripts, with each point in the graph representing the preceding year's arithmetical mean of each content class (Y-axis) over time (X-axis). We created one graph for each category to serve as a basis for a qualitative time series analysis.

In Figure 28 the effect of the one-year moving average results in a smoothing effect of the graph. There are still changes to recognize and the peaks are easier to locate.

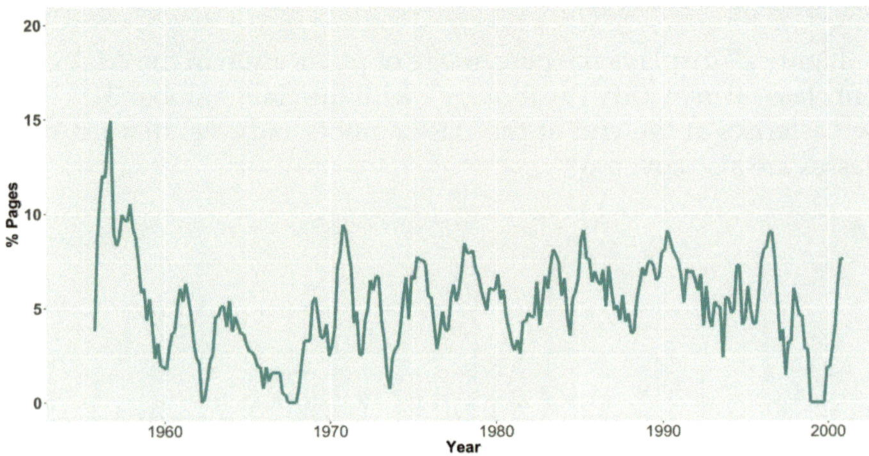

Figure 28: Application of a one-year moving average

If we now add the strong positive decisions of Communist Party Congresses as solid purple lines and meetings of the Central Committee of the Communist Party as solid blue lines, we get the next Figure 29.

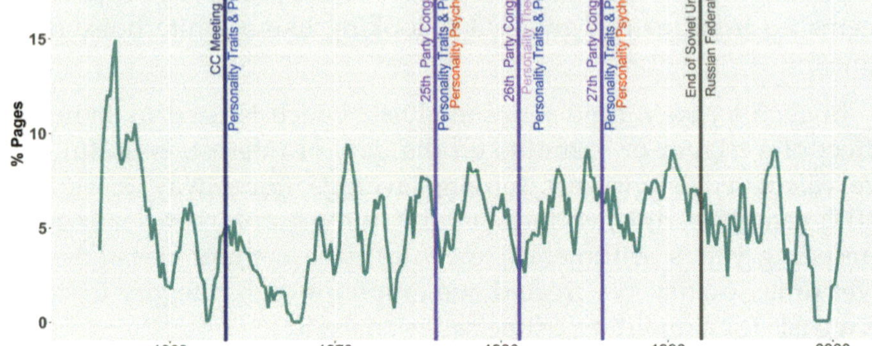

Figure 29: Strong positive Communist Party decisions on Personality Psychology

Now we can identify some connections of Communist Party decisions to peaks in the graphical representation of the article percentage in the APA content class "Personality Psychology.

In the next figure 30 we add also the purple dashed lines, which are indicators of relatively weak positive decisions (allocated less than 1 page in the respective article) at the given Communist Party decisions.

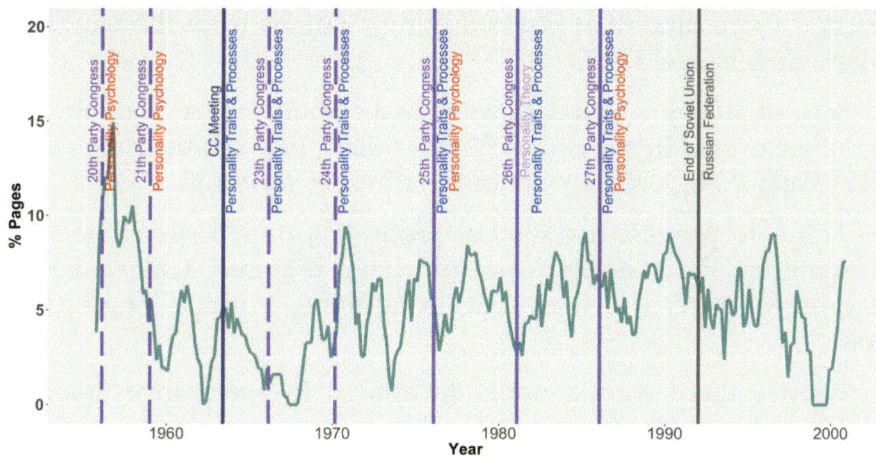

Figure 30: All Communist Party decisions on "Personality Psychology"

Now there are more connections to recognize, which relate peaks of the graph with party decisions. Unfortunately, not every peak is related to a Communist Party decision and not every Communist Party decision lead to a peak in the article percentage frequency.

Therefore, some qualifying conditions have to be noted for the interpretation of the graphs presented in the next chapter:

First, it must be pointed out that our experiences with translating the article titles indicated that there was a great quality gap between those articles dealing theoretically with the party's goals and aiming to confirm the ideas of Marx, Engels, and Lenin on the one hand and those reporting experimentally collected data or statistics on the other. The first kind of the articles could be published shortly before the party congresses or shortly after. Krementsov (1997) pointed out the existence of a strong relation between researchers and their "sponsors" in the Communist Party. So many authors knew in advance which tasks would be on the agenda in an upcoming Communist Party congress.

The second type of articles were typically published some years after the first type of articles, as data collection and interpretation needed more time than just collecting citations from the works of Marx, Engels, and Lenin.

Second, of course, all data were, to a certain degree, confounded by other events in the Soviet Union taking place at the time other than Party Congresses or Central Committee meetings.

Third, to provide a statistical proof of a time series caused by Communist Party decisions would have required a much larger number of party decisions and, in addition, a much longer time span.

Fourth, there were random factors as in every investigation, which had no systematic relationship to the events under investigation.

Based on these limitations, we present the percentages of pages in Voprosy Psychologii taken up for publishing papers in a particular subfield of psychology (defined according to the APA psychology content classification), as a function of year, political leader, and important meeting. We tried to find a link between the political milieu and the fluctuation of the percentage of pages used for publishing papers in a specific branch of psychology over the time period covered in this study.

Limitations And Caveats Of Our Research

One issue concerning our data is the fact that the years after 1977 saw a trend of new psychological journals being launched, and this trend has since increased. At present it is hard to count them, as many institutions publish their own journals, and this is not controlled by the government. Today there may be 30 or more journals. So, from 1977 onwards, we did not capture 100% of the scientific articles in the field of Soviet and Russian psychology. In addition, there were other non-psychology disciplines in which psychologists published and still publish their findings and theories, including pedagogy, physiology, philosophy, psychiatry, biology, and more, which are not covered in our research.

To sum up, we did not capture all articles in the field of Soviet and Russian psychology, but we included 100% of the contributions in Voprosy Psychologii (7,049 in total), which was the most relevant Soviet psychology journal and the only one from 1955 to 1976.

As in previous eras, the Communist Party and its First Secretary decided on the tasks of the sciences and how they should improve socialist society, i.e. the way of life in Socialist and Communist countries. In the following chapter, we examine these tasks for psychology, which were published as articles in Voprosy Psychologii, and their effect on the academic output of this journal. Each of the Communist Party's decisions pertained one to several APA content classes, which is reflected over the times in the graphs.

How To Read The Following Graphs

Each graph in the following chapter represents the one-year moving average (Shumway & Stoffer, 2017) of the percentage of pages of a specific subfield of psychology (APA content class) occupied as described in the previous paragraph.

The **blue vertical bars** show the time of the **Communist Party Central Committee** meeting that made a positive decision concerning this content class.

The solid **purple bars** show relatively strong positive decisions of a given **Communist Party Congress**, which was explained in the journal article with 1 page or more. The purple dashed bars are indicators of relatively weak positive decisions (allocated less than 1 page in the respective article) at the given Communist Party congress.

The **gray bars** indicate the changes of the party's First Secretary or Russian President.

Impact Of The Communist Party's Tasks For Psychology In The Post-Stalin Era (1955–2000)

Creation Of The New Soviet Man

One of the very early tasks of the Bolshevik Communist Party was the creation of a new kind of human being, one who would act like a "cog in the wheel" (Gerovitch, 2007). As Yasnitsky (2016) said, "... one of the key tasks of the post-revolutionary era was utopian 'remolding of man,' the creation of a new type of people, who will master their nature and uncover the yet unknown potential of human beings. These ideas were grounded in the pervasive post-revolutionary belief in the possibility of virtually unlimited personal growth and an active, creative attitude to the world" (p. 5). Petrovsky characterized the tasks of education and psychology like this: "School, education, upbringing of a new man - this is an area which was coined by the intensive search for new methods and practices of pedagogical and psychological work. From the first months of the revolution on, Soviet psychologists were active in the pedagogical search" (citation according to Petrovsky, 1967, p. 21).

Yasnitsky (2016) says: "The role psychology was to play in this social transformation was very special and highly important. Psychology was to find the means for the normative remolding of the 'old man' of the capitalist past and educating the 'new man' of Communism. These methods would be subsequently implemented in large-scale social projects and would lead to the creation of the improved and advanced people of the future. ... Therefore, it was not abstract, theoretical interest, but the urgent demands of social practice that determined the rapid development of applied psychoneurological disciplines grounded in the actual concrete tasks of the establishment of a new society" (Yasnitsky, 2016, p. 6). According-ing to Gao (2019), the same goals were pursued in the other Communist countries like China.

In the early years after the revolution, there was the belief that only the new socialist society would automatically create new humans, with the pressure of capitalist society disappearing. As we know now, this was not the case, and it was Stalin who reestablished law and order from 1929 on. There were also attempts to produce "new humans" by controlled breeding in the 1920s (Mocek, 2002), but those attempts failed.

From the October revolution onwards, there was also a belief that the person best suited for a position in the party, the administration, and the army should get the job. This was different from the selection by class criterion adopted in Tsarist times, which favored nobilities. According to the new selection criterion, the supposedly best way to recruit personnel was by using tests. So from 1920 on, multiple tests were developed, and many people were hired for administering those tests. However, the massive use of tests by inadequately trained people led to unsatisfactory results. In addition, as Stalin learned, the people he wanted to see in leading positions – descendants of rural and industrial workers – were not the ones who typically obtained the highest scores; it was rather the descendants of middle-class parents and land owners who usually scored high in the tests.

This led to the Pedology decree in 1936, in which all tests and statistical interpretation of facts and scientific results were abandoned. Now the main selection criterion for party, administration, and government positions was ancestry again, but in a reverse way compared to Tsarist times. The parents of a candidate had to be rural or industrial workers. Psychology became a kind of neglected child of pedagogics (Petrovsky, 2000).

Now that statistics were banned, the new criterion for the accuracy of research was its agreement with the ideas and writings of Marx, Engels, and Lenin. This was the so-called required "partisanship of science," one of the traditional principles of dialectic materialism (London, 1952).

From 1948 on, all psychology had to be based on the principle of conditioned reflexes (according to Pavlov), and education had to follow this approach.

However, from 1955 on, psychology gradually returned to the methods of the 1920s and hence the theories of psychologists such as Vygotsky, Rubinshtein, and Luria were now in fashion again. Nevertheless, the aim of creating the "new human" remained a very important goal of psychology and education. Even Gorbachev believed it was not the socialist system that was to blame for the economic problems the Soviet Union was facing but the people who still were not of the type the system needed (Taubman, 2017).

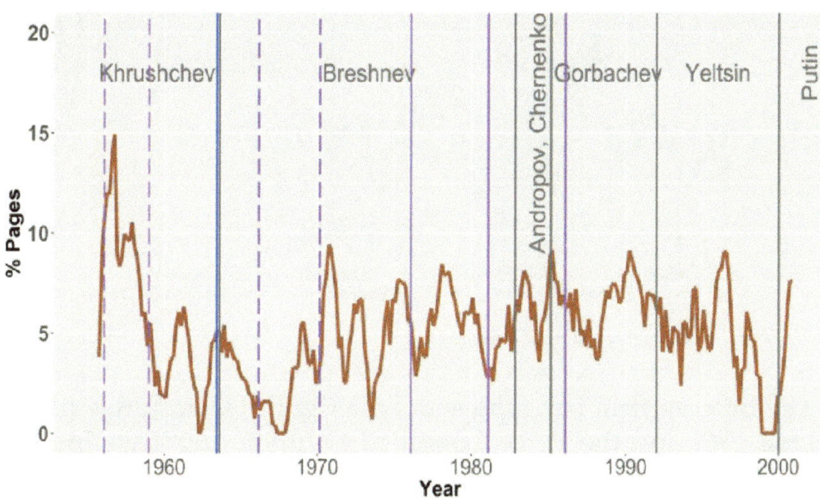

Figure 31: Personality Psychology

"According to the 'Moral Code of the Builder of Communism,' a model Soviet citizen was expected to be an active member of society and to take 'an uncompromising attitude' toward any injustice or insincerity. At the same time, an exemplary citizen was supposed to have 'a strong sense of social duty.'" (Decisions of the 22nd Party Congress of the CPSU, cited according Taubman, 2017)

This main goal affected the APA content classes (see Figure 26) Personality Psychology, Developmental Psychology, Social Psychology, and Educational psychology.

Figure 31 shows the one-year moving average of the page percentage in all the issues from 1955 until the end of 2000 in Personality psychology. In order to facilitate reader's the interpretation of the graphs, we provide the following explaining notes:

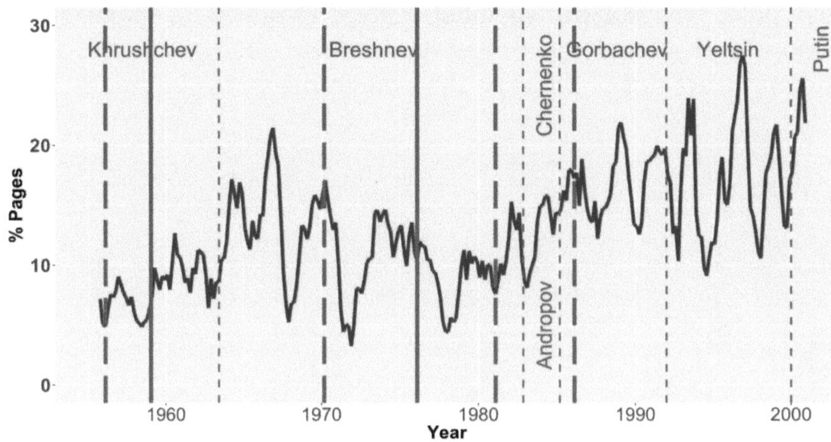

Figure 32: Developmental Psychology

The blue vertical bar represents the Central Committee meeting in June 1963 and the Party Congresses with strong tasks in February 1976, 1981, and 1986 are marked with solid purple bars. There were also Party Congresses with minor emphasis on Personality Psychology in 1956, 1958, 1966, and 1970, marked with dashed purple bars.

We can notice a high start with nearly 15% of the published pages, which could be due to the scientific conference in July 1955, dedicated to the "Theory of Set" (Uznadze, 1966). An additional effect might be due to the minor decision of the Communist Party

Congress in February 1956. Until 1961 there was a decline of this class' publication page percentage. After the 1959 party congress there was a minor peak in 1961. At the Central Committee meeting in 1963, there was a minor peak, too. From 1966 onwards, there was a steady increase which peaked near the Party congress meeting in 1970. Until 1986 it appeared that every Party congress had an additional impact on Personality psychology. In the following years there were no party decisions any more and the variations were due to other factors. As mentioned, there were other events influencing the effects of page percentages in a content class. In addition, there were random factors that did not show any systematic effects. To address all the factors in detail would be beyond the scope of this chapter.

Figure 32 displays the moving average in Developmental Psychology. Again, it appears that there was a connection between the graph's peaks and the party's decisions. The 1959 Party congress made an emphatic decision on the tasks of Developmental Psychol-

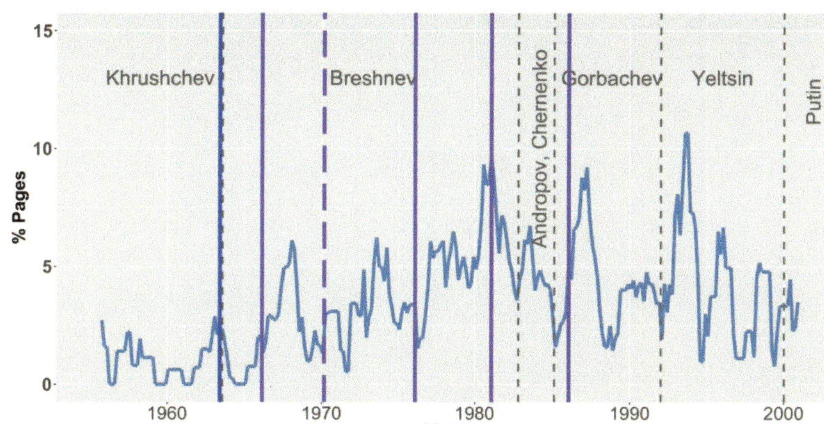

Figure 33: Social Psychology

ogy. Together with the preceding Party congress in 1956 (with minor decisions) there was a steady increase of publication percentages in this area of psychology until 1967. There was a minor peak at the Party congress in 1970. After the 1976 Party congress meeting, there was a steady increase again, which continued even after the End of Soviet Union in 1991.

It is remarkable to notice that the peaks of Personality Psychology coincided with the lows of Developmental Psychology and vice versa. This indicates the high priority of the "molding of the new personality" for the Communist Party.

Social Psychology gives a slightly different picture: Prior to the Central Committee meeting in 1963, social psychology did not exist in the Soviet Union. Instead, the theories of Marx, Engels, and Lenin provided the right answers for this field. At the Central Committee meeting in 1963 (it was the time of de-Stalinization), the party learned from research in Western countries, especially from the US, of the benefits of social psychology, which indeed began to

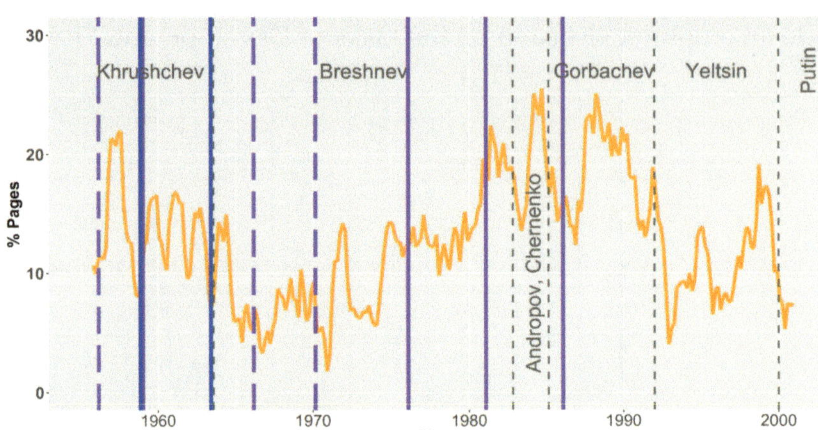

Figure 34: Educational Psychology

emerge in the Soviet Union at this point (see Fig. 33).

The results for educational psychology (Figure 34) are much more complicated. There were multiple requirements for educational psychology, mainly related to adapting the curriculum to the new party demands. We can see a high level of this content class in the 1950s, when there was a high demand for closer links between theory and practice, and a renewed increase in the 1970s until the end of the Soviet Union, when the government sought to improve the quality of school education, especially in science, for the first time. From 1985 onwards, perestroika was a topic in school education. The demand for the creation of the new human seemed to be a minor part of the development in this content class.

Improvement In Quality, Output, And Organization In Industrial And Agricultural Production

There was a second task of the Soviet Communist Party which appeared in nearly all reports of the party congresses and the CC meetings in 1958 and 1963 which was: improving the quality of agricultural food processing and industrial production, increasing

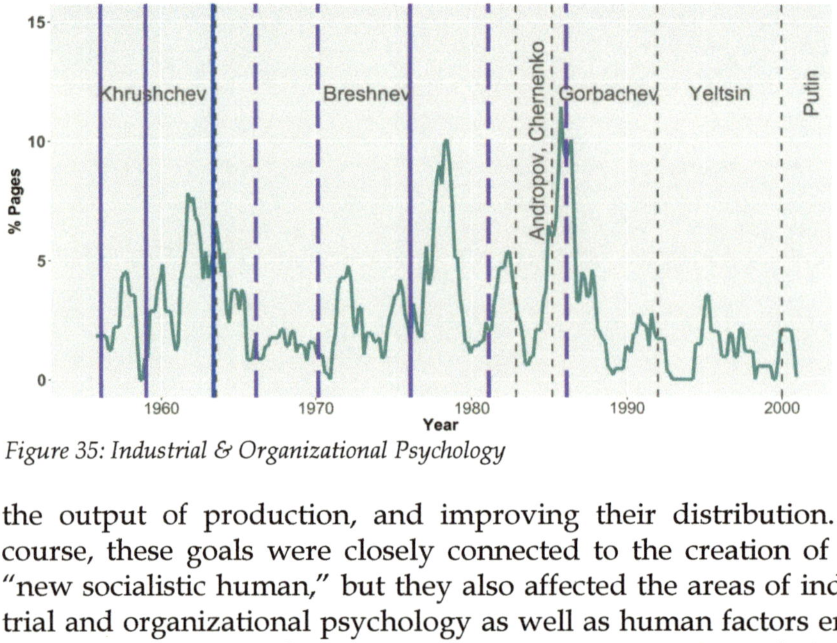

Figure 35: Industrial & Organizational Psychology

the output of production, and improving their distribution. Of course, these goals were closely connected to the creation of the "new socialistic human," but they also affected the areas of industrial and organizational psychology as well as human factors engineering (main part of Engineering & Environmental Psychology). These reports of the party decisions indicated the problems in these areas and the need for improvement, but a Communist way by which to achieve the goals was not specified.

Figure 35 shows the development of Industrial and Organizational Psychology from 1955 to 2000 and the Communist Party's decisions concerning this field. The publication peaks seem to be related to the party's decisions. There are two trends in matching party decisions with the relatively early publications in this field: First, the tight connection of the researcher's sponsors in the Communist Party, which gave them the chance to publish the desired papers shortly after the decisions or even in time. Second, there were articles dealing theoretically with the party's aims, confirming findings related to the publications of Marx, Engels, and Lenin (instead of reports of experimentally collected data).

Likewise, in Engineering and Environmental Psychology (Figure 36), the peaks of the published articles in this area appear close to the time of the party's decisions. However, there was a remarkable decrease in publications after 1975, which nearly coincided with the peaks in the related area Industrial & Organizational Psychology. Perhaps the focus on human factors engineering had shifted to management, personnel selection, and training.

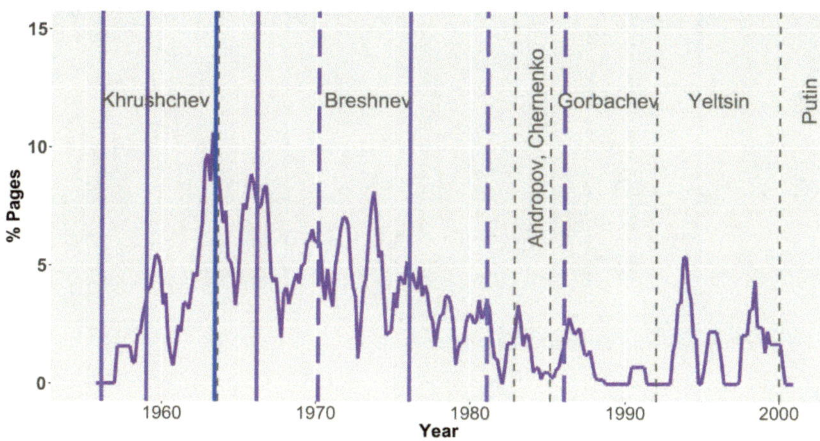

Figure 36: Engineering & Environmental Psychology

Diagnosis And Treatment Of Physical And Psychic Disorders

The diagnosis of physical and psychic disorders was addressed at several Communist Party congresses, although treatment was not a major issue at these meetings. Psychic disorders were mainly treated by sending patients to forced labor camps, where they had to do hard work in the company of criminal delinquents and dissidents.

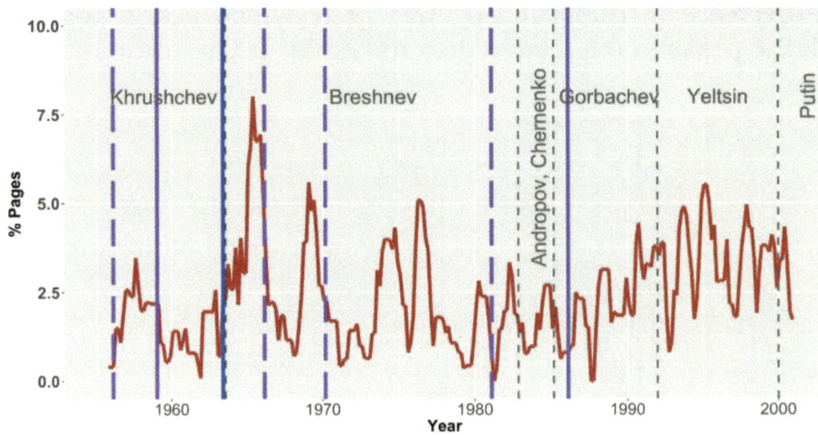

Figure 37: Physical and Psychic Disorders

As stated before, the psychologists' tasks consisted mainly of diagnostic issues and expert opinions (Lauterbach, 1976). The treatment was the task of the psychiatrists, who were closely connected with the jurisdiction.

This situation changed a little bit for the better in the late 1970s, when Brezhnev's power weakened and the Communist Party started to care more for the welfare of their members than for political aims (Lauterbach, 1976).

There were several party decisions on the diagnosis of Physical and Psychic Disorders (see Figure 37). Publications in this area

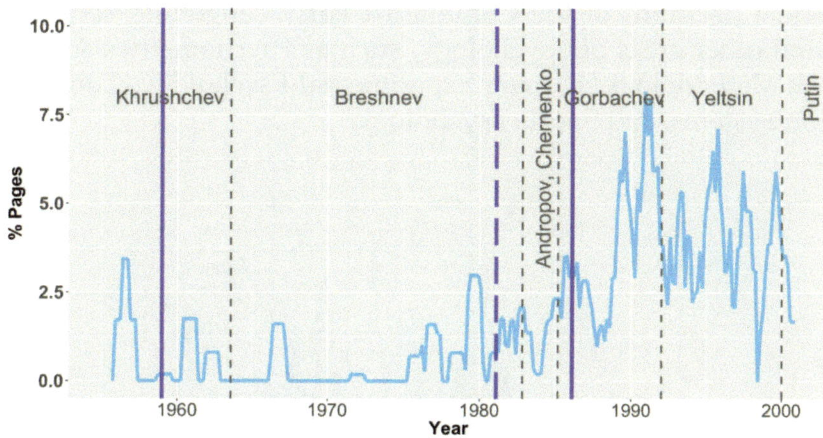

Figure 38: Health & Mental Health Treatment & Prevention

peaked after a Central Committee Meeting in 1963. Health & Mental Health Treatment & Prevention (Figure 38) shows a different picture: The Communist Party Congress in 1959 made a decision to improve this area, but it had nearly no effect on the publications, which stayed close to zero until the mid-1970s.

When party control weakened in the mid-1970s, the percentage of Treatment & Prevention articles increased, presumably reflecting the increase of the interests of psychologists and the needs of psychologically sick people for treatment. As we can see in some other categories, from 1975 on, the contents of the journal more closely resembled those of Western psychological journals according to the experience of the authors.

Other Goals Set By The Communist Party

Some decisions of the Communist Party congresses concerned several other areas of psychology, such as Psychometrics & Statistics & Methodology, Human Experimental Psychology, Communication Systems, and Sports Psychology.

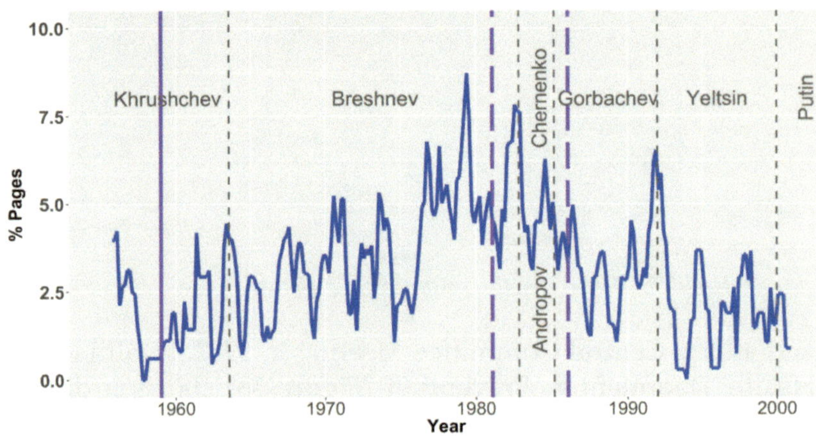

Figure 39: Statistics and Research Methods

Since 1937, the use of all statistics and tests in psychology and education had been banned in the Soviet Union. The criteria for assessing theories and research findings were strictly based on their consistency with the principles in the publications of Marx, Engels, and Lenin. This was called the required "partisanship of science", that means, the priority of Marxist-Leninist theories in science. As Stalin said in 1948, the Party had its own position in scientific questions, and this position was true by definition (Schattenberg, 2017).

The attitudes towards statistics and research methods in psychology changed after Stalin's death. Thus, the 21st Communist Party congress in 1959 made a decision to improve the application of statistics and research methods. Figure 39 shows a steady increase in publications in this content class, which decreased after

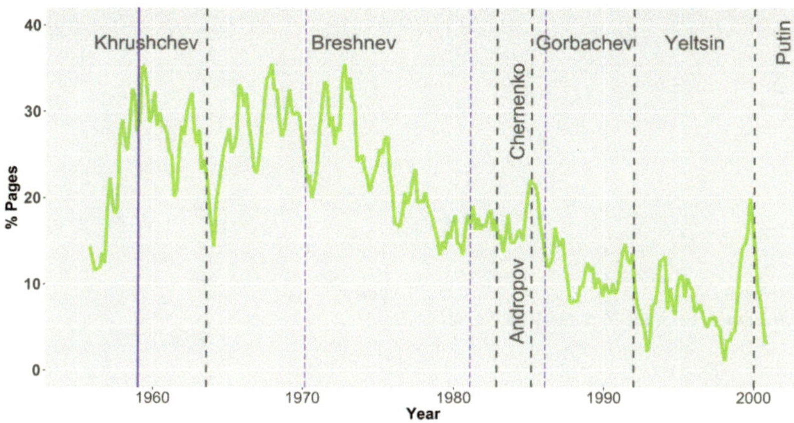

Figure 40: Human Experimental Psychology

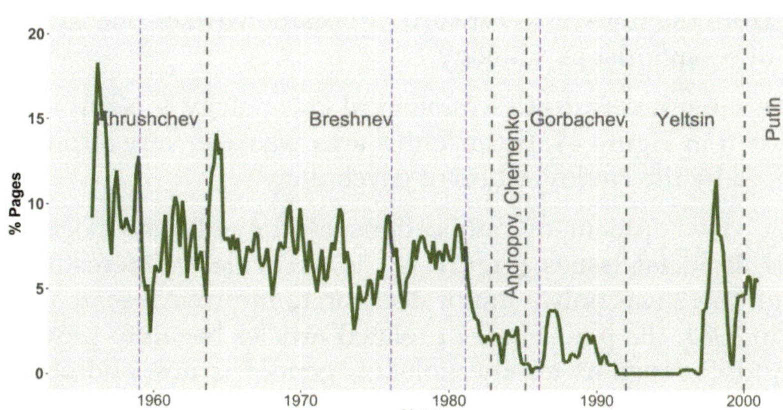

Figure 41: Physiological Psychology & Neuroscience

1985 during perestroika and in the founding years of the Russian Federation due to changing priorities in psychology.

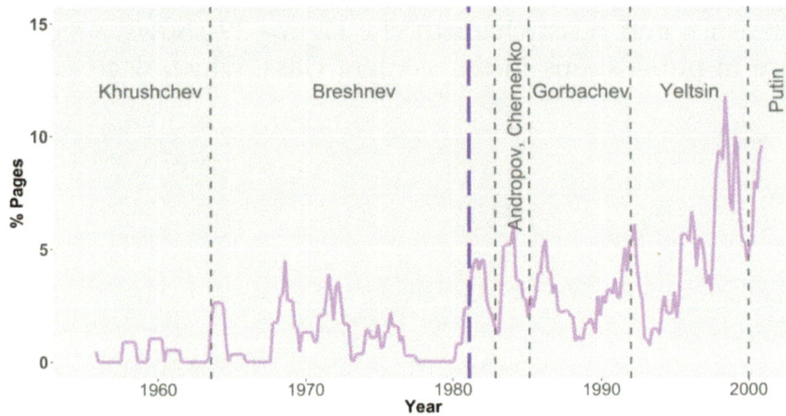

Figure 42: Social Processes & Social Issues

Human Experimental Psychology was a traditionally strong area in Soviet psychology, partly due to the Pavlovization of psychology since 1948. In 1959, there was a decision to further strengthen this area. Figure 40 illustrates the positive effect. However, from the mid-1970s onward, it became weaker due to the decline of behaviorist psychology.

The same was true of Physiological Psychology & Neuroscience, as shown in Figure 41, because this area was also very strongly influenced by the Pavlovization of psychology.

The development of publications of content class Social Processes & Social Issues (Figure 42) is particularly interesting. Although there was only a minor decision to improve research in this area in 1981, the percentage of related articles began to grow, covering topics such as family violence, gender issues, and abuse of drugs and alcohol even before this decision of the 26th party congress. This was one more indicator of the growing freedom of science and the decreasing influence of the Communist Party.

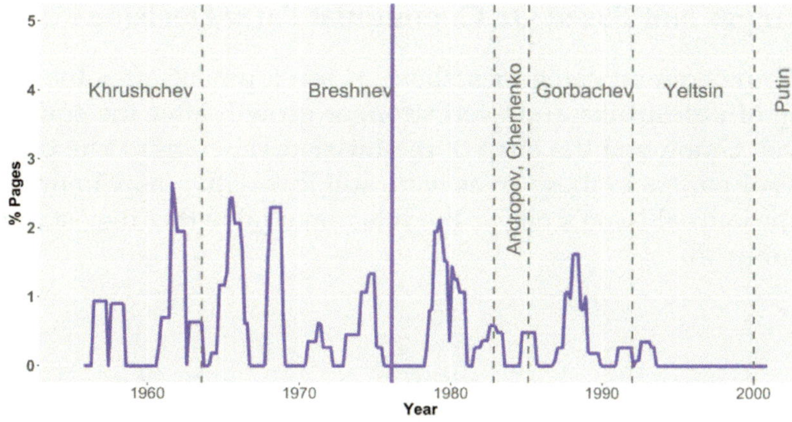

Figure 43: Sports Psychology

In Sports Psychology (see Figure 43) and Communication Systems (see Figure 44), we see no relationship between party congress decisions and the percentage of publications. These decisions were made after 1975 and might also be an indicator of weakened party influence on science.

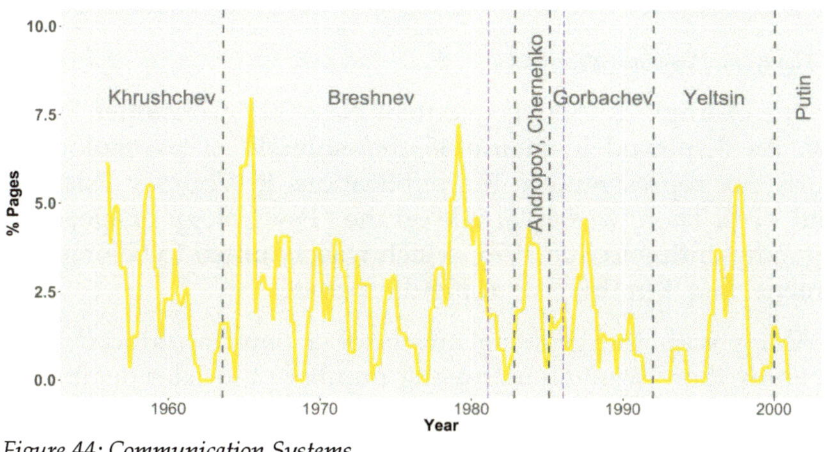

Figure 44: Communication Systems

Changes Not Based On Communist Party Decisions

Some content categories show a development after the beginning of perestroika and even stronger growth after the end of the Soviet Union and the birth of the Russian Federation. The percentages of papers in these areas were still low with a maximum of 5% of the journal issue's pages, but it is remarkable that they started to occur at all.

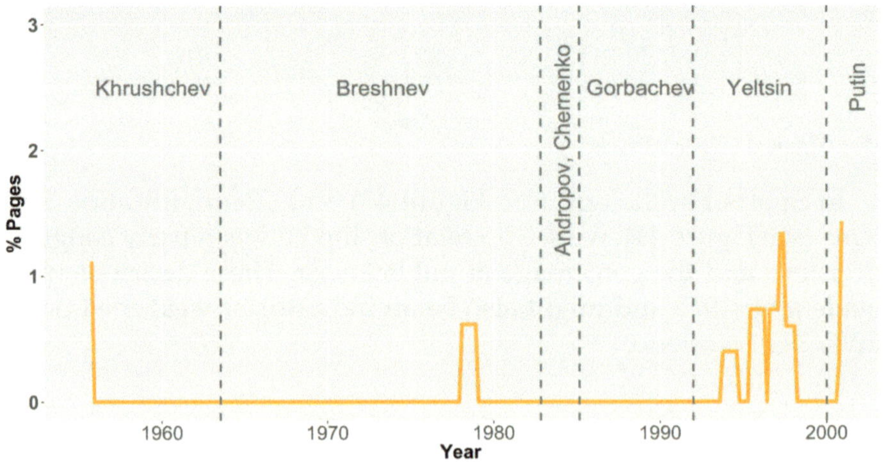

Figure 45: Consumer Psychology

One example is the APA content class 39** Consumer Psychology. As displayed in Figure 45 this subfield of psychology was nearly not represented in the publications in Voprosy Psichologii until 1990. There was an article on the "Psychology of speech propagandist influence" in 1978, which was assumed to belong to this content class. But this was a singular paper.

Along with the growing effects of a new introduced market economy there were an increasing number of articles dealing with problems of the market economy. In total six articles dealt with subjects like "Content analysis of lonely hearts advertisements", "Dynamics of the emotional reaction of Russian customers to advertising", "Consumer's economic behavior and the theory of activity", "Gender images in advertisements".

Two subfields of main APA content classes deserve special attention: Environmental Issues & Attitudes (4070), subclass of 40** Engineering & Environmental Psychology, and Special & Remedial Education (357*), subclass of Educational Psychology (35**).

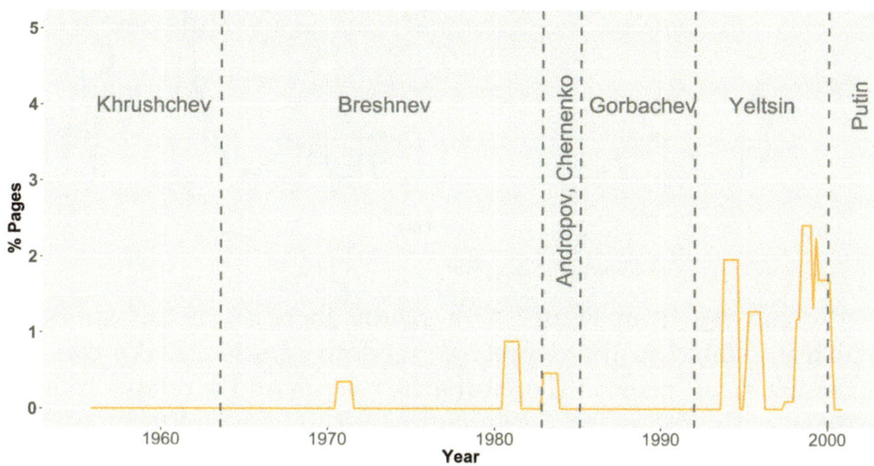

Figure 46: Environmental Issues & Attitudes

Out of the 12 articles in "Environmental Issues & Attitudes" in Figure 46 9 appeared since 1990. They address problems like " Perception of radiation risks by specialists in the field of nuclear energy and nonspecialists", "Nature protection and it's perception by humans", and "Psychological aspects of school ecology education".

The other subfield 357* "Special & Remedial Education" includes subsubfield 3575 "Gifted & Talented Education". As Educational Psychology was always a content class representing nearly 13% of the article pages in the investigated 45 years, it is not surprising, counting 46 articles in total. 10 articles have been published in 30 years before 1985, 10 between 1985 and 1989, and the majority of 26 between 1990 and 2000.

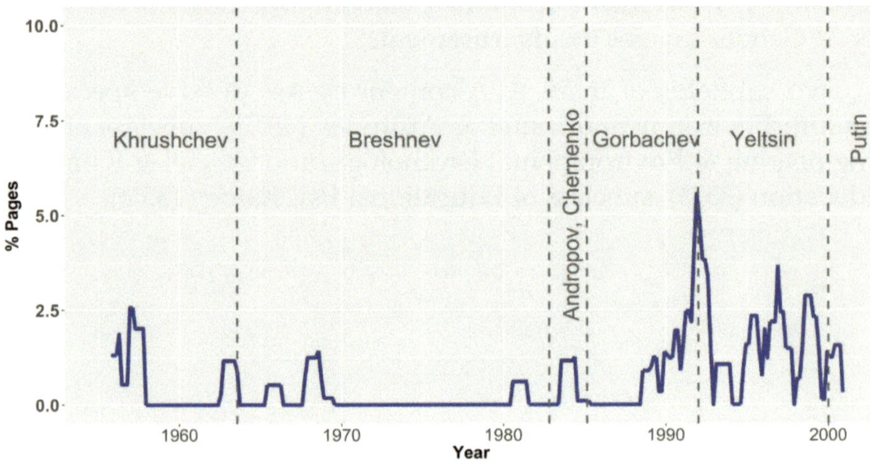

Figure 47: Special & Remedial Education

We can see from Figure 47 a steady increase of this subfield, which is related to the degree of freedom of science. We can observe the steep peak in the Gorbachev years and a relatively high plateau in the following years, compared to 25 years from 1955 to 1980 of low amount of article percentages of this content subclass.

This is a strong indicator that the freedom to pursue topics in science generates research questions that are relevant to society.

Discussion

Our bibliometric research indicates a general interrelation between the percentage of publications in a specific APA content class and the Communist Party's decisions at the Central Committee meetings and the party congresses. From the late 1970s onward, Russian psychologists worked more and more on topics relevant to society, to the people of the Soviet Union, and to those who needed psychological treatment.

It is very interesting to recognize that with the beginning of perestroika, and especially after the foundation of the Russian Federation, new areas of interest evolved for psychologists, which had been neglected before: Health & Mental Health Treatment & Prevention, Social Processes & Social Issues, Special & Remedial Education, Consumer Psychology, and Environmental Issues & Attitudes. This development shows that freedom of scientific research led to the discovery of problems in society and the desire to find solutions for these problems.

Conclusions And Further Study

Psychology and psychologists in the post-Stalin era enjoyed relatively more freedom than in the eras before, except for the first years following the Bolshevik October revolution in 1917. Nevertheless, during most of the period surveyed in this study psychological research and practice continued to be supervised and controlled by the Communist Party and Communist Party members, who acted as "sponsors" for the psychological institutes and their staff (Krementsov, 1997).

To obtain a more complete picture of the changes in psychological research topics in the post-Stalin era of the Soviet Union, we plan to include more journals in the areas of psychology, pedagogy, physiology, and philosophy in our future investigations. Since many of these journal articles will hardly fit into the APA content classes system, we plan to perform a "text-mining" analysis for the article titles. We believe that the rise and fall of expression frequencies (e.g., "new man", "personality") or in name frequencies (e.g., "Pavlov", "Vygotsky") across the publication titles will give us hints about the preferred trends over time in Soviet and Russian psychology research.

Table 1: Articles concerning Communist Party's Decisions on Psychology published in the journal *Voprosy Psichologii*

Author	Year	Issue	Pages	Title in English Translation
Academy of Pedagogical Sciences	1956	2	3-7	The 20th Congress of the CPSU and the Tasks of Psychological Science
Academy of Pedagogical Sciences	1958	6	27-32	Decisions of the November Plenum of the Central Committee of the CPSU and Tasks of Psychology]
Smirnov, A.A.	1959	5	7-28	Tasks of Psychology in the Light of the Decisions of the 21st Congress of the CPSU
Academy of Pedagogical Sciences	1963	4	3-6	The June Plenum of the CC of CPSU and the Tasks of Psychology
Academy of Pedagogical Sciences	1966	3	3-9	The 23rd Congress of the CPSU and the Tasks of the Pschological Science
Academy of Pedagogical Sciences	1971	1	3-6	Soviet Psychology On the Eve of the 24th Congress of the CPSU
Academy of Pedagogical Sciences	1976	1	3-9	To the 25th Congress of the CPSU
Academy of Pedagogical Sciences	1976	2	3-8	The 25th Congress of the CPSU and the Tasks of Soviet Psychology
Feldstein, D.I.	1976	3	3-15	The 25th Congrss of the CPSU and psychological Problems of the Communist Education
Lomov, B.F.	1976	6	9-19	Decisions of the 25th Congress of the CPSU and the Tasks of the Psychological Science in the Fight for the Increase of the Effectivity and Quality
Melnikov, V.M.	1978	4	3-8	Actual Problems of Sports Psychology in the Light of the 25th Congress of the CPSU

Author	Year	Issue	Pages	Title in English Translation
Academy of Pedagogical Sciences	1981	5	5-15	Towards the 26th Congress of the CPSU
Bodalev, A.A.	1981	2	5-10	The 26th Congress of the CPSU and the Tasks of the Psychological Science
Academy of Pedagogical Sciences	1981	4	5-11	The 26th Party Congress of CPSU on the tasks for general education and the psychological-pedagogical science
Apollonov, V.L.; Sluzky, E.G.	1981	4	184-186	The 26th Congress of the CPSU and Actual Problems of Social Psychology
Parygin, G.D.	1981	6	5-12	The 26th Congress of the CPSU and Contemporary Problems of Social Psychology
Zotova, O.I.	1982	1	162	Problems of Socialistic competition in the Light of the decisions of the 26th Party Congress of CPSU
Academy of Pedagogical Sciences	1986	1	5-10	Towards the 27th Party Congress of CPSU
Matyushkin, A.M.	1986	5	5-17	The main tasks for the psychological research in the light of the decisions of the 27th Party Congress of CPSU

References

Altrichter, H. (2013). *Kleine Geschichte der Sowjetunion 1917-1991. [Short History of the Soviet Union.]* (4.). Ch.H. Beck.

Artemyeva, O. A. (2015). *Социально-психологическая детерминация развития российской психологии в первой половине XX столетия [Social and Psychological Determination of the Development of Russian Psychology in the First Half of the Twentieth Century]*. Cogito-Center.

Bauer, R. A. (1959). *The New Man in Soviet Psychology*. Harvard University Press & Oxford University Press.

Bratus, B. S. (1998). *Русская, советская, российская психология [Russian and Soviet Psychology]*. Biblioteka Shkolnogo Psikhologiya.

Budilova, E. . (1975). *Philosphische Probleme in der Sowjetischen Psychologie. [Philosphical Problems in the Soviet Psychology.]*. VEB Deutscher Verlag der Wissenschaften.

Bushkovitch, P. (2012). *A Concise History of Russia (Cambridge Concise Histories)* (1st ed.). Cambridge University Press.

Colton, T. J. (2008). *Yeltsin - A Life*. Basic Books.

Covington, C., & Wharton, B. (2015). *Sabina Spielrein. Forgotten Pioneer of Psychoanalysis* (2nd ed.). Routledge.

Davydov, V. V. (1982). Психологическая наука в СССР и Школа [Soviet psychology and school education]. *Voprosy Psychologii*, *6*, 21–34.

Dickinger, C. (2001). *Franz Joseph I. Die Entmythisierung [Franz Joseph I. The Demythization]*. Carl Ueberreuter.

Ehrsam, M. (1985). Zur Entwicklung einer marxistischen Persoenlichkeitspsychologie in der UdSSR. [On the Development of a Marxist Psychology of Personality in the USSR.]. In E. Gorny & H. Knopf (Eds.), *Zur Aktualitaet sowjetischer psychologischer Theoriekonzepte* (pp. 6–16). Universitaet Halle-Wittenberg, Sektion Erziehungswissenschaften, Wissenschaftsbereich Psychologie.

Eichhorn, H., & Stern, G. (1977). Zur Geschichte der Psychotherapie in Russland und der Sowjetunion. [On the history of psychotherapy in Russia and in the Soviet Union.]. *Psychiatrie, Neurologie Und Medizinische Psychologie*, *29*(10), 577–586.

Field, A., Miles, J., & Field, Z. (2012). *Discovering Statistics Using R* (1st ed.). Sage Publications, Inc.

Fritsche, C. (1980). Zur Frühgeschichte der Internationalen Kongresse für Psychologie. Ein Überblick [On the early history of the international psychological conferences. A survey]. *Wissenschaftliche Zeitschrift Der Karl-Marx-Universität Leipzig - Gesellschafts- Und Sprachwissenschaftliche Reihe*, *29*(2), 167–171.

Gao, Z. (2019). Forging Marxist psychology in China's Cold War geopolitics, 1949-1965. *History of Psychology*, *22*(4), 309–327.

Gerovitch, S. (2007). "New Soviet Man" Inside Machine: Human Engineering, Spacecraft Design, and the Construction of Communism. *OSIRIS*, *22*, 135–157.

Graham, L. R. (1972). *Science and Philosophy in the Soviet Union*. Vintage Books.

Hamann, B. (2010). *Kronprinz Rudolf. Ein Leben [Crown Prince Rudolf. A Life]* (4th ed.). Piper.

Hodnett, G. (1974). *Resolutions and Decisions of the Communist Party of the Soviet Union, Volume 4. The Khrushchev Years: 1953-1964.* (R. H. McNeal (ed.)). University of Toronto Press.

Hyman, L. (2017). The soviet psychologists and the path to international psychology. In J. Renn (Ed.), *The globalization of knowledge in history* (p. 854). Max Planck Institute.

Joravsky, D. (1986). *The Lysenko Affair*. The University of Chicago Press.

Keiler, P. (1988). Die Anfangsetappe der sowjetischen Psychologie und der kulturhistorische Ansatz der Wygotski-Schule [The initial stage of Soviet psychology and the cultural-historical approach of the Vygotsky school of thought]. In N. Kruse & M. Ramme (Eds.), *Hamburger Ringvorlesung Kritische Psychologie. Wissenschaftskritik, Kategorien, Anwendungsgebiete* (pp. 37–81). Ergebnisse-Verlag.

Kesselring, J. (2011). Vladimir Mikhailovic Bekhterev (1857-1927): Strange circumstances surrounding the death of the great Russian neurologist. *European Neurology, 66*(1), 14–17.

Khlevniuk, O. V. (2015). *Stalin - New Biography of a Dictator* (1.). Yale University Press.

Knoell, H. D., & Jou, J. (2018). Soviet and Russian Psychology From 1950-2000 (From Stalin to Putin). *APA Annual Meeting*, 379.

Kozulin, A. (1984). *Psychology in Utopia - Toward a Social History of Soviet Psychology*. The MIT Press.

Krementsov, N. (1997). *Stalinist Science*. Princeton University Press.

Kussmann, T. (1974). *Sowjetische Psychlogie: Auf der Suche nach der Methode [Soviet psychology: in search of the method]*. Hans Huber Verlag.

Lauterbach, W. (1976). Die Lage der klinischen Psychologie in der Sowjetunion [The state of clinical psychology in the Soviet Union]. *Psychologische Rundschau*, *27*(27), 225–236.

Lewin, M. (2016). *The Soviet century* (2nd ed.). Versobooks.

London, I. D. (1949). A historical survey of psychology in the Soviet Union. *Psychological Bulletin*, *46*(4), 241–277. https://doi.org/10.1037/h0063507

London, I. D. (1952). Sowjetische Psychologie und Psychiatrie [Soviet psychology and psychiatry]. *Ost-Probleme*, *4*(16), 495–501.

Massie, R. K. . (2013). *Nicholas and Alexandra: The Tragic, Compelling Story of the Last Tsar and his Family* (Kindle ebo). Head of Zeus Ltd.

McLeish, J. (1975). *Soviet Psychology: History, Theory, Content*. Methuen & Co Ltd.

Myers, S. L. (2015). *The new Tsar - The Rise and Reign of Vladimir Putin*. Simon & Schuster.

Perrez, M., & Krampen, G. (2015). Comorbidity in Clinical Psychology Research 1980-2014: Publications Trends and Topics in the Anglo-American versus the German-Speaking Countries. In *ZPID Science Information Online* (Vol. 15, Issue 6).

Petrovsky, A. W. (1967). *История Советской Психологии. [History of Soviet Psychology.]*.

Petrovsky, A. W. (2000). *Психология в России [Psychology in Russia]*. Verlag der Universität der russischen Bildungs-Akademie.

Poole, R. A. (2002). Moscow Psychological Society. *Routledge Encyclopedia of Philosophy*.

Richebaecher, S. (2008). *Sabina Spielein. Eine fast grausame Liebe zur Wissenschaft [Sanina Spielrein. An Almost Cruel Love of Science]* (1st ed.). btb.

Rubinshtein, S. L. (1971). *Grundlagen der allgemeinen Psychologie [Basics of general psychology]*. Verlag Volk und Wissen.

Rueting, T. (2002). *Pavlov und der neue Mensch [Pavlov and the New Man]* (1st ed.). R. Oldenbourg Verlag.

Schattenberg, S. (2017). *Leonid Breschnew [Leonid Brezhnev]*. Böhlau Verlag.

Schwartz, D. V. (1982). *Resolutions and Decisions of the Communist Party of the Soviet Union, Volume 5. The Brezhnev Years: 1964-1981.* (R. H. McNeal (ed.)). University of Toronto Press.

Sebestyen, V. (2017). *Lenin the Dictator*. Weidenfeld & Nicolson.

Shub, D. (1962). *Lenin - eine Biographie [Lenin - a Biography]*. Limes Verlag.

Shumway, R. H., & Stoffer, D. S. (2017). *Time Series Analysis and Its Applications. With R Examples.* (4th ed.). Springer.

Taubman, W. (2003). *Khrushchev - The Man and his Era.* W.W.Norton & Co.

Taubman, W. (2017). *Gorbachev - His Life and Times*. Simon & Schuster.

Thielen, M. (1984). *Sowjetische Psychologie und Marxismus [Soviet Psychology and Marxism]* (1st ed.). Campus Verlag.

Todes, D. P. (2015). *Ivan Pavlov. A Russian Life in Science*. Oxford University Press.

Tucker, R. C. (1974). *Stalin as a Revolutionary - A Study in History and Parsonality, 1879-1929.* Norton & Co., Inc.

Tucker, R. C. (1990). *Stalin in Power - The Revolution from Above, 1928-1941.*

Tuleya, L. G. (2007). *Thesaurus of psychological index terms.*

Volkov, V. A., Gvozdetsky, V. L., Orel, V. M., & Urmantsheyev, M. A. (1988). *Наука и техника СССР 1917-1987: Хроника [Science and Technology of the USSR 1917-1987: Chronicle]* (I. istorii estestvoznaniia i tekhniki. Akademiia nauk SSSR (ed.)). Nauka.

von Rauch, G., & Geierhos, W. (1990). *Geschichte der Sowjetunion [History of the Soviet Union]* (8th ed.). Alfred Kröner Verlag.

Yasnitsky, A. (2016). The Archetype of Soviet Psychology. From the Stalinism of the 1930s to the "Stalinist science" of our time. In A. Yasnitsky & R. Van der Veer (Eds.), *Revisionist revolution in Vygotsky studies* (pp. 3–26). Taylor & Francis.

Zubok, V. M. (2009). *A Failed Empire* (2.). The University of North Carolina Press.